Tackling Tomorrow Today

Volume One: Futuristics: Looking Ahead

Volume Two: America: Moving Ahead

Volume Three: Getting Personal: Staying Ahead

Volume Four: Moving Along: Far Ahead

Courtesy of The Venus Project
Designed by Jacque Fresco and Roxanne Meadows

Tackling Tomorrow Today

Volume Four
Moving Along: Far Ahead

Edited by Arthur B. Shostak, Ph.D.
Emeritus Professor of Sociology,
Department of Culture and Communication
Drexel University, Philadelphia, PA 19104

Philadelphia

CHELSEA HOUSE PUBLISHERS

VP, NEW PRODUCT DEVELOPMENT Sally Cheney
DIRECTOR OF PRODUCTION Kim Shinners
CREATIVE MANAGER Takeshi Takahashi
MANUFACTURING MANAGER Diann Grasse

Staff for TACKLING TOMORROW TODAY

EXECUTIVE EDITOR Lee Marcott
EDITOR Christian Green
PRODUCTION EDITOR Noelle Nardone
PHOTO EDITOR Sarah Bloom
SERIES AND COVER DESIGNER Takeshi Takahashi
LAYOUT EJB Publishing Services

©2005 by Chelsea House Publishers,
a subsidiary of Haights Cross Communications.
All rights reserved. Printed and bound in the United States of America.

A Haights Cross Communications ↖ Company ®

http://www.chelseahouse.com

First Printing

9 8 7 6 5 4 3 2 1

Library of Congress Cataloging-in-Publication Data

Tackling tomorrow today / edited by Arthur B. Shostak.
    p. cm.
  Includes bibliographical references and index.
  ISBN 0-7910-8401-9 (v. 1) -- ISBN 0-7910-8402-7 (v. 2) -- ISBN 0-
7910-8403-5 (v. 3) -- ISBN 0-7910-8404-3 (v. 4) 1. Twenty-first
century--Forecasts. 2. Technology and civilization. I. Shostak,
Arthur B.
  CB161.T33 2004
  303.49'09'05--dc22
                                            2004016198

All links and web addresses were checked and verified to be correct at the time
of publication. Because of the dynamic nature of the web, some addresses and
links may have changed since publication and may no longer be valid.

■

*Dedicated to forecasters,
prominent and unsung alike,
who help us see further,
imagine more,
prepare better
and
savor life's extraordinary possibilities.*

■

There is nothing permanent except change.
—Heraclitus

■

*To keep our faces toward change
and behave like free spirits
in the presence of fate
is strength undefeatable.*
—Helen Keller

ACKNOWLEDGMENTS

Sixteen high school students from six states and the District of Columbia volunteered to critique well over sixty candidate essays and help me choose fifty-eight for the four volumes in this series. Their cogent and insightful feedback (266 brief reviews) can be found at the rear of each volume, and it makes clear my considerable debt to them: Amelia Adams, Mike Antonelli, Erin Bauerle, Patricia Marie Borrell, Andrew Crandall, Alex Dale, Tom Dunn, Rebecca Henderson, Mara James, Sarah Konner, Ginger Lemon, Kelly Ramirez, Dalea Reichgott, Benjamin Samuels, Brittany Tracy, and Jessica Varzaly. Special thanks go to Alex, Dalea, Mike, and others for survey responses, and for sheer output alone, to Tom, Jessica, Benjamin, Mara, Patricia, Ginger, Andrew, and Alex.

Plainly, much appreciation is owed the forty-three writers of the series' fifty-eight original essays; busy people who took time to share creative ideas and earnest feelings about our choices in making probable, possible, preferable, and preventable futures.

Several contributors (Glenn, Jeff, Joe, Josh, Marilyn, Nat, Roger, Sohail, and Tom) commented usefully on the essays of others. John Smart secured remarkable artwork for his two essays from Cris Dornaus. Marvin Cetron, Nat Irvin, Mel Konner, Robert Merikangas, and Patrick Salsbury ably adapted essays. Ann Coombs provided special research material of great value. While they did not write essays, Daniel Shostak did provide insightful discussion questions, as did

Nada Khader. Jacque Fresco and Roxanne Meadows shared their extraordinary artwork.

Many whose ideas are not aired directly in the book nevertheless made a vital contribution. Stevi Baggert, Connie Cordovilla, Judith Czigler, Alexander Friedlander, Thad McKenna, Adrienne Redd, and Emily Thorne helped recruit high school volunteers. Todd R. Grube identified editorial cartoons of high quality. Peggy Dominy, an Information Services Librarian at Drexel University, found hard-to-locate missing data. And, of course, there were many others whom I trust will forgive my regrettable memory lapse.

As before in the case of five books I edited in 2003/2004 on 9/11 and the Iraq War (DEFEATING TERRORISM/DEVELOPING DREAMS), the staff of Chelsea House did an especially fine job meeting some rather complex challenges, with special thanks going to the series editor, Christian Green.

For more than a quarter of a century, my wife, Lynn Seng, has contributed ideas of great value, reviews of keen insight, and support without which I would accomplish far less. Her belief in this project, and her love and smile, make all the difference.

Finally, I would like to acknowledge YOUR unique contribution, for it is ultimately only as you—and other readers—ponder and act on the book's many ideas that this volume can help us craft a world tomorrow that increasingly honors us all.

Table of Contents

ACKNOWLEDGMENTS 6

INTRODUCTION 13

PART ONE: INFORMATION TECHNOLOGY—TOMORROW 17

Essay One
A Digital Day
Jan Amkreutz 19

Essay Two
Future Heroes 2035: The Big Picture
John Smart 30

**PART TWO: BIOSCIENCES/BIOTECHNOLOGY
—TOMORROW** 44

Essay Three
Our Genetics Century: Wow!
Graham T.T. Molitor 46

Essay Four
Letters to Unborn Daughters: Exploring
the Implications of Genetic Engineering
Sarah Stephen 52

Essay Five
It's Twelve O'Clock, and I Know Exactly Where
My Youngsters Are
John Cashman 59

Essay Six
Biotech and Food: Making a Finer Future
Graham T.T. Molitor 66

Essay Seven
Biology in Your Future: Prepare to Become
Somebody New
Melvin Konner, Ph.D. 72

PART THREE: NANOTECHNOLOGY—TOMORROW 83

Essay Eight
Nanotechnology: Big Revolution with Small Things
Jim Pinto 85

Essay Nine
... and the Bubblegum Pops: Nanotech vs. Capitalism
Glenn Hough 95

PART FOUR: SPACE—TOMORROW 100

Essay Ten
Space: Teenagers and the Far Out
Jeff Krukin 102

Essay Eleven
Debate: Space and Our Near Future
Jeff Krukin 115

Essay Twelve
Why Should We Send Humans to Mars?
Thomas Gangale 124

Essay Thirteen
A Mars Colony: C096
John A. Blackwell, Phil Gyford, Glenn Hough,
Alexandra Montgomery, and Dana Wilkerson-Wyche 132

EPILOGUE

On Using Futuristics 147

Essay Fourteen

The Future of Us: How Can We Finally Become
Really and Truly Human?
Melvin Konner, Ph.D. 149

APPENDIX

Student Feedback 161
Abstracts from *Future Survey* 175
Annotated Bibliography (O–Z) 189

NOTES ON CONTRIBUTORS 202

INDEX 206

Introduction

We won't just experience 100 years of progress
in the twenty-first century— it will be
more like 20,000 years of progress.
—James John Bell, "Exploring the 'Singularity,'"
The Futurist, May–June 2003

In this closing volume of our four book series—*TACKLING TOMORROW TODAY*—we use fourteen essays to underline an idea that courses through and binds together all of the books: namely, that it is possible to prepare for at least some of what is coming, take charge of a bit of the rest, and build confidence and faith that we can handle whatever else remains.

Two essays in Part One delve into leading challenges involving information technology, arguably the greatest enabler at present of profound and rapid change in the sciences, the arts, and whatever. The first takes us into a day in the life, this time, of an adult businessman a decade or more ahead, when life is lived in a wireless reality of thinking machines. The second helps us better appreciate what the arrival of machines with minds superior to ours might mean, approaching a time ahead called the Singularity, as interpreted for us by a sixteen-year-old high school student in 2035.

Five essays in Part Two explore the pros and cons of likely advances in biology, chemistry, medicine, and drugs. Promising breakthroughs are heralded, even as vexing questions and alarms are underlined: For example, do we really want our babies custom-crafted before birth in a laboratory? Do we really want our children to carry a biochip in their bodies so we can track them anywhere? Do we really think the gains of genetically modified foods outweigh the alleged risks?

Part Three focuses on the world of the tiny, the *very* tiny—as in one-billionth of a meter, or one 80,000[th] of the width of a human hair. Known as nanotechnology, and detailed in the first essay, this frontier in the sciences is already a worldwide multi-billion-dollar sensation. Its supporters hail it as the Third Great Revolution (after Agriculture and Industry) and contend that even if only half of its potential is realized in the next decade, that is enough to radically change almost everything now familiar about reality. Capitalism, for one, as made clear by the second essay, is challenged (threatened?) as never before.

Four essays in the next part would have us soar into space, only to be jolted to a halt by far-reaching disputes. Our essayists debate whether we should get our own house (Earth) in order before shouldering the costs of going fully into space? Whether we should send people or robots to Mars? Whether private industry should play a larger (for-profit) role? And finally, what might a Mars colony resemble, and is it the stuff that dreams are made of? (If not, how might you improve it?)

A capstone essay weaves together much that has come before, not just in this fourteen-essay volume, but in all fifty-eight essays in the series. The writer explores lessons he draws from about 7 million years of evolution and uses these insights to suggest a still-finer future that we are invited to help co-create.—Editor

REFERENCES

Boulter, Michael. *Extinction: Evolution and the End of Man*. New York: Columbia University Press, 2002. A challenging forecast of life on Earth after humans have become extinct, even though other forms of life survive.

Cooper, Richard N., and Richard Layard, eds. *What the Future Holds: Insights from Social Science*. Cambridge, Mass.: MIT Press, 2002. Nine thoughtful essays wrestle with such questions as why have past forecasts been so poor; how many humans will there be; will there be enough energy; and how will climate change affect our lives.

Dixon, Dougal, and John Adams. *The Future Is Wild*. Toronto: Firefly, 2003. This book fast-forwards to Earth millions of years from now, long after humans are extinct. A revolutionary and fascinating approach to evolution.

Foundation for the Future. *The Future Human: The Next Thousand Years*. Bellevue, Wash.: FFTF, 2003. Nine far-sighted intellectuals engage in a fascinating and illuminating dialogue.

Lightman, Alex, with William Rojas. *Brave New Unwired World: The Digital Big Bang and the Infinite Internet*. New York: John Wiley & Sons, 2002. Explores various aspects of a wireless world, and forecasts convergence of technologies in a single elegant and user-friendly personal communicator.

Sterling, Bruce. *Tomorrow Now: Envisioning the Next 50 Years*. New York: Random House, 2002. Ultra-creative exploration of links among technology, culture, human history, and the future.

Part One

INFORMATION TECHNOLOGY— TOMORROW

The lesson of the last three decades is that nobody can drive to the future on cruise control.
—Rowan Gibson,
Rethinking the Future

As is well known, computer power doubles about every eighteen months, and its cost halves. The Internet continues to double every two years, and we race pell-mell into a new world of ubiquitous computing, both wireless and increasingly affordable.

To be sure, challenges abound from an onslaught of spam, eBay fraud, identity theft, and ever more vicious viruses. We take heart, however, from the good done by Web sites available to the confused and needy; from high-quality online dating services; from the ability of once-timid souls to speak up (anonymity is possible); and from the demonstration by Howard Dean's candidacy of the power of Web users to shake the foundation.

Essay One envisions what a day might be like in 2020 or sooner when we live in a fishbowl of wireless smart equipment capable of a dialogue with us. Essay Two pushes the matter further and brings us into a world where the equipment is—

arguably—much smarter than we are (though our poetry may remain superior).

Enthusiasts can hardly wait, while critics dismiss the scenario (called the Singularity) as unlikely at best and highly undesirable at worse. It is not too early for you to begin arriving at an informed position of your own.—Editor

■ Essay One ■

A DIGITAL DAY

Jan Amkreutz

President, Digital CrossRoads

Prologue. Imagined scenes. Snapshots from an ordinary day, not too far in the digital future. That is the story of this essay—a story, however, with a question mark. Is this imagined day just a composite of "gee whiz" technologies already happening or at least starting to happen today? Certainly, they might be fully realized by 2020 or even much sooner. Or does the scene reveal an emerging new world behind a deceptive curtain of playful gadgets?

6:15 A.M.: Briefing. My house is trying to wake me up—the bed performs a gentle massage to "Nature's Dreams," one of those tunes that bring the sounds of nature to soothe the human soul. Usually, this gets me out of bed quickly, but not today. Since I'm still not moving, the bed switches to Andrea Bocelli's *Con Te Partiro.* I get the message, and with it, a new sense of urgency.

The bedside lamp explains: "The house is already up and ready to go, Jay. You have a breakfast meeting at 7:30 with Angie Rapallo at the Echo Café." It projects Angie's face on the ceiling. Darn. I forgot about that. *"Where are my notes?"* I ask, adding, *"And get my car warmed up. Don't forget to switch on the seat heaters."*

"Already done, and the notes are in your shirt," the bed answers. It's too late for the treadmill, so I have to get the morning news from the bedroom mirror. I stumble to the bathroom, to consciousness, and a new sense of time.

Back in the bedroom to dress, I ask the mirror for a selection of morning news bites of sounds and images I prefer. While it compares today's news with yesterday's, it stays in "mirror" mode; time enough to shave. (Thank God for comparing news. My own digital news agent compares for content and assembles only what's new since yesterday. If I want to, I'll get a summary of yesterday's news to bring me up to date.)

"Want to hear your schedule for the day?" "*Sure.*" "After your breakfast meeting, no more physical appointments for the rest of the day, until meeting your wife for dinner. She comes in on the shuttle at 7 P.M. and meets you at 8 P.M. at the Rose Garden. She will call you later. At 10:30 A.M., you will meet Carlson from New York, Villeneuve from Paris, and Lee from Hong Kong.

"Lee only reports on the research findings, so her mind will not participate; otherwise, it is expected to be a highly interactive meeting, so your personal presence is expected. Mind-presence only, though. You could do it from the car." "*Good. Then I might finally look at those real estate properties, remember?*"

"That's exactly what I suggest. Your next appointment is at S–."

"*Ouch!*" I zoom in on the mirror to study the result of a clumsy shave. "*Oh well, let's have the news, anyway.*" Better get showered. As an ISN reporter comments on the progress of the Mars colony, I think back to the old days of television, when hundreds of channels showed the same programs, over and over again. Oh, some still exist today. Interestingly, those who finally confessed to their biased reporting found a new and valuable niche.

ISN (Inter Society News) is one of the major feeds to the Meta-News-Web, formerly CNN. MNW was one of the first fully interactive TV "networks," which integrated the old Internet with traditional TV. They transmit multiple, linked programs that each participant can tailor to personal preferences. The result is that I can receive opposing viewpoints on a certain subject, no matter what channel it comes from, look at different subjects with the same bias, or get a high-level overview of different perspectives.

Anyway, I finish my shower and dress. My shirt is of the

newest variety, lightweight and with much higher functionality. Its built-in 360-degree airbag gives protection in all kinds of unforeseeable situations. It monitors my vital signs and gives me advanced warning for such things as heart attacks or strokes. The newest ones even warn me in case of certain bacterial or viral infections.

This shirt maintains my DNA signature and produces a new one, whenever a fresh one is required. (Using somebody else's DNA is a felony and prosecuted stringently. Many places do require an instant test using your hand as the source of DNA. Iris-scans proved too easy to fake, especially with new organic iris replacements or alterations using laser surgery.) All my medical information is attached to my DT (my "digital twin"). My DT only provides information to authorized people.

My digital counterpart is my extended memory. It stores all information pertaining to me, from my DNA signature to the odometer reading of my car; from a detailed three-dimensional model of my house to the balance in my checking account; from my childhood pictures to my favorite music today. In the old days, doctors, insurance companies, supermarket chains, online retailers, or the government owned information about me. Not any longer. I own it: It is my DT. I have to authorize any use of it and never transfer ownership.

My shirt starts doing its morning routine, as it sorts through the notes sent by the bed. I dictated those notes last night on my tablet-PC, which doubles as an alarm clock and a remote control for the house. The shirt transcribes my notes, so I can use them at my breakfast meeting. Finally, it measures my vital signs and transmits the data to the doctor's digital twin.

I unplug my shoes from the charger and put them on. I had to replace them recently, because of new GPS technology with much greater accuracy. These days you need reliable wireless communication, because you want your family members to know where you are. With this technology, you can find your way around places, because office buildings, shopping malls, and fun parks recognize you, know where you are, and give you directions. Of course, the subscription fee to these services is cheaper if you agree to watch or listen to commercials.

The annoying thing is that you have to buy new shoes when a new technology becomes available. No "plug and play" for shoes these days. The shoes themselves are the expensive part; that's how they make their money—antenna, receiver, and power supply, all in the shoes.

My mocha is ready, chocolate and all, prepared by the newest model of the "Star'O'bucks" espresso machine. I grab the coffee, take a sip and ... I find the chocolate is missing! The machine explains: "Your shirt told me that your blood sugar is too high this morning, therefore no chocolate, as per your own instructions. By the way, it's decaf. Your blood pressure, remember?" Oh well, can't blame DT for taking care of me. The machine's voice adds: "The chocolate is getting low, anyway. I'll have the car remind you."

On my way to the door, the microwave stops me in my tracks: "Don't forget about tomorrow's date." When the microwave's door displays my wife's smiling face in bright colors, I remember: Tomorrow is her birthday.

7:00 A.M.: On the Road. The morning is cold, the car is nice and warm, and the batteries are charged. "Want me to turn the house, J.J.?" the car asks. *"Sure, and I forgot to open the curtains. The house is still in manual mode."* "Done. I turned it to auto, so it will move to keep the windows on the sunny side. Don't forget to pick up chocolate. Want me to find a place that has your brand in stock, or you want it delivered? The refrigerator answers the door anyway." *"I'll pick it up at Charlene's. Those folks always crack me up. I like them, and I like their good humor. Just remind me when I'm close."*

The Echo Café is small, warm, and never too full. It has a free Net-connection of only two hundred megabytes per second, but it will do for the meeting. Miss Rapallo is an ambassador for the "Global Lifestyle Society," one of those new Planetary Organizations, commonly called POs. They provide their members with a full array of personal, profes-sional, and social services. Much debate is going on, since many traditional nations still refuse to relinquish power to other, more global organizations.

POs provide all the traditional insurances, like life, home, car, health, liability, and so on. The newer ones go far beyond that by providing services that used to be exclusively performed by the government. They provide such things as unemployment benefits, disability benefits, and the usual law-enforcement and security protection services, which they contract out to local governments. Membership assures you of these services no matter where you live in the world.

GLS, the one represented by Miss Rapallo, has fifty governments under contract, but your choice is not limited to those countries. If you live in one or more of the fifty under contract, however, or if you commute between them for your work, the services provided are more elaborate. Your membership fees cover your taxes (as we used to call them), which in turn the organization pays to governments for their services. Your social security money does not go to governments but, at your option, stays under your control or goes to a choice of affiliated financial services firms.

When I park the car in front of the Echo Café, the café's charger connects itself to the car. "Angie Rapallo is going to be a few minutes late. I just talked to her car. She is ten kilometers away. She knows you're here." I instruct the car to get ready for the real estate trip later on.

8:00 A.M.: Planetary Breakfast. I'm on my second cup of coffee. Angie had given me the story of her organization, which was very similar to the story of the society I belong to. "I just wanted to explain to you that we offer all the services that you currently enjoy. Please allow me to get to where we differentiate ourselves from your organization." *"Please go ahead, Angie."*

"Our philosophy can be summed up in two short sentences." She takes a display from her purse and rolls it out. The two sentences emerge in bright colors from a background image of my favorite village in the Black Forest of Germany: *freedom of lifestyle, relationships create substance.* She continues to explain the services of the organization while my mind wanders through my memories of that village.

"We realize relationships are at the core of any lifestyle, and our services help you connect to people with similar interests around the globe, whether these interests are of a religious, scientific, cultural, or other nature. We help you to establish and maintain exchanges. These inter-relationships create the substance of what we call citizenship. That is why we call our members *citizens*."

Interesting. Forty years ago, that would have sounded eerie. Over the years, the planet, and its incredible diversity, had become so visible through digeality; it gave you a different perspective. The last war in the '00s had further accelerated the awareness of this great diversity. It was the start of a mind-shift. Now, with people on Mars, my perspective on the planet Earth had changed. Now, I consider it my home, if you will, especially since my last trip to the Moon.

9:15 A.M.: Don't lose your shirt. I had gone through my notes, asking Angie all the questions I had prepared, and we agreed to be in contact in a couple of days. My shirt had wired the slides she had shown to my home-server. While I drive away, the car asks: "Were you happy with the services in the café? You forgot the tip."

On the way out, I verbally confirmed the bill at the door; the shirt confirmed my identity and paid the bill. "Losing your shirt" has a new meaning these days. I was supposed to add the amount of the tip either vocally or by entering it on a writing-pad. I forgot. *"Just send two Globals; with my apologies."*

"Done. You want to listen to the memo for your upcoming meeting?" *"Sure."*

Unlike many other people, I'm old-fashioned when it comes to implants. Because of a severe hearing problem that kept getting worse, my wife persuaded me to get cochlear implants. They double as a phone and a microphone. I'm connected to the world all the time, unless I switch it off just by telling it to shut up. The car simply forwards the memo for the meeting to my ear as I listen. The meeting is "mind-presence only": my

DT will be present in digital or holographic form in the meeting room. I'll be connected through my cochlear antenna and react verbally.

My DT will actually speak my words and mimic my gestures. It's neat, because it gives the people in the room a sense of presence. For a while, hackers managed to breach such meetings by creating some "interesting" disruptions. Anyway, newer technologies, whereby I have full digital presence are still experimental. With new technology, my DT is fully present and the physical me sees through its eyes and so on. With all of that new technology, true physical meetings have become a real "treat," and obtain a new significance.

The new global positioning system guides me to the first property I want to see. My wife had studied the place in digeality, in three dimensions—all the details. Except the trees. The property was shown in the summer, with lush flora hiding many of its features. Now, in winter, the trees are bare and allow a view of the "naked" property. She asked me to have a look at it. The owner of the property indicated that it would be for sale three years from now, when he planned to retire. Sometimes you could make a good deal, if the owner had the certainty of selling the house *today* and could stay for the three years.

"*Damn!*" I did it. I saw the house in the last instance, hit the brakes, and the snow on the road did the rest. "I can't pick up the signals from the road, J.J. Too much snow. I'm sorry." Sure enough, I hit a sign that reads, "Dead End," just as the car says: "No road-signal." I wish them all to some place far away: the sign, car, digital, house, all of it.

"I've done a preliminary estimate," the car says soothingly. "The total is $1,200, the insurance company approves, and here are your options for body shops. You're lucky, the frame is OK, and the fender repaired itself. You can drive. Just give me a visual check. Oh, your right blinker doesn't work, but you just need to replace the LED. Here is the closest supplier."

12:50 P.M.: Precious issues. The meeting had gone well. I had not been able to attend, because I had driven my car straight to the body shop. The parts came from across town and had been delivered by the time I got to the shop, which didn't have a Net connection suitable for my meeting. Therefore, I had given some last-minute instructions to my digital twin to participate without me.

Over the years, new software, called the "semantic janitor," had built an extensive profile of the language I use in such meetings and the meaning of the concepts I try to convey. Using the same software, my DT had done the same for the people I meet frequently, so that a basis for mutual under-standing had been established.

The subject of the meeting was our low retention rate of highly talented people. The proposal on the table was a risky one for the company, but then, the times required innovative solutions. We had negotiated a two-sided contract with one of the major "Knowledge Co-ops" (KCs) that was based on the continuity requirement of our company to retain talent and the long-term interests of the KCs' members.

This particular KC had relationships with large communities of members in Russia, Sweden, the former East Germany, India, and the Northwest Territories of Canada. New communities were in the process of being established in the Southwest United States and Mexico. Since the United States became officially bilingual in 2015, the mobility of Spanish-speaking people between North and South America has been increasing. Russia and northern Canada have provided climate-enhanced environments that greatly improved the living conditions in their remote territories.

In return for a "social security retainer" that we pay to the KC, we have guaranteed availability of the talent we need, while being able to select that talent on a project basis as needed. The retainer paid for social security and pension bene-fits of the individual members and the continuity gaps of the KC.

In addition, we agreed to participate in the organization's "mobility and lifestyle" program, meaning that individuals and their families can choose a new residence once a year, if they desire. This gives our employees the possibility to participate in the program as well. The KC had excellent relationships with schools in many regions and partially financed the continuous "knowledge renewal" of the schools' online and offline teachers.

"I'll be home early; I managed to get an early *glide* home with Jessica. Want to have an early dinner at six?" my wife says. *"Great. I'll see you at six at the Rose."*

6:05 P.M.: A new concept. A lazy afternoon. After I had replayed the meeting, the car was ready to go at 3 P.M. I proceeded with my real estate "tour" and got a birthday present for my wife, making sure she could return it. If she says something like "how original," I would know: return was imminent. Thank God some things never change.

"Darling, this is the most original present you ever came up with!" See what I mean. The thing was as good as back in the store. *"Wonderful, I'm glad you like it. How did your strategy session go?"* My wife works as a "lifestyle counselor," a new profession that grew out of the former real estate and travel agent professions.

"Would you like to change your mind about your selection?" the waiter asks. Both of our DTs had looked at the menu earlier in the day, and preselected our meals based on our preference, what we had last time, what we said about it, the specials of the day, and so forth. The DTs kept themselves up to date on all of the menus around town, except the few restaurants that didn't maintain digital ones.

"I'm OK," my wife says. "Are you?" I look at my selection on the table in front of me. *"Fine with me."*

"I want to tell you about my glide. Jessica has one of the newer ones. They are truly whisper-quiet, no wonder they are called 'glides.' Flight control had to divert us to a parking lot nearby, because the Rose's parking is too small for Jessie's glide. Hers is one of the new van-type models that can accommodate up to twenty people."

"Tell me about the meeting." "Well, I submitted my proposal to localize our services. Our competitors are catching up with

the services we offer. As you know, the inference engine that your company sold us gave us the advantage of providing better knowledge to our customers. Your engine had better semantic description and cross-referencing capabilities.

"Now, the competition is catching up. More important, our clients expect a more personalized approach. They want counselors to understand their own personal interpretations of 'a good neighborhood,' a 'good place to live,' or 'a preferred way of living.'"

"Our new product does exactly that." "Darling, I know. But the product has to sell itself; I can only do so much for you." Indeed, some things never change.

(Digital connections are necessary, but not sufficient, to do business. Our new software allows users to define their own links between the knowledge provided in cyberspace and the meaning the inference engine attaches to the knowledge. That way, especially for clients attaching a portion of their DTs to our engine, the counselors can "program" the software to derive personal interpretations.)

"The essence of my proposed concept is a human approach. Only human counselors can establish the final link between the interpretations of the client and your software. We already introduced a 'couleur locale,' because our advice to clients depends on the *local* circumstances of the regions of the planet under consideration. With my new concept, we want to add a 'couleur humaine,' if you will."

"Couleur humaine?" "Yes, darling. You see, we live in the age of female intuition, remember? The guys used to call it 'holistic thinking' or 'systems approach.' Mind research has demonstrated the innate ability of the mind to form coherent holistic views of situations. Until your software can do the same, we'll have to trust our human 'instincts.'"

"Funny thing, our mind. The moment our software catches up, the mind seems to travel beyond the capability of the software." "Our minds focus on renditions, darling, not the underlying algorithm. Your analysts and programmers have to deal with algorithm that produces new renditions. We as counselors

simply take the new renditions into account and produce new knowledge from them, capiche?"

"Keeps us in business, though. How's your dinner?"

"Excellent, as usual. Cannot wait to get home, though, and drop my shoes. I'm tired of them."

8:00 P.M.: Dropping the other shoe. Finally, we could drop our shoes. These days that was one of the most important moments of the day, since our connection to the outside world was in the shoes. Now we were disconnected, and the evening was ours.

The ride home had been uneventful. As we approached the water, the house had already turned toward the passageway that led to the road. My car was an old-fashioned "auto-mobile," and could only handle old-fashioned roads and calm waters. We had bought the house a few years ago to escape the hustle of suburbia.

The house was part of a generous subdivision, just half a mile off the coast. A part of the ocean had been tamed by a system of submerged barriers that rose and lowered with the tide, just like the house itself. In cold weather, we lowered the house into the warm coastal waters to save energy and get away from the noise of the nearby coast.

At dawn, the house would rise again and find the Sun, follow it, and absorb the Sun's energy to power our lives for the coming day.

Epilogue. A digital day in some imagined future. How should we feel about it? Is it frightening? Amusing? Promising? Fascinating? Pure nonsense? Something that is too far away to worry about? Science fiction? *Who* cares…?

WE must care—for whatever feelings the "digital day" produce in our minds, this digital reality is of historic evolutionary importance, and it can change not only *what* we see, *but also how we see it.* And that changes the choices we can make. If the curtain of technology is hiding a new world, we should know our choices, because the choices we make today are the contours of that new tomorrow.

■ Essay Two ■

FUTURE HEROES 2035: THE BIG PICTURE

John Smart
President, Institute for the Study of Accelerating Change

I'm Dev, and I go to Fremont High, in Rolling Hills, CA, US of A. The year is 2035. Most people would label me a futurist, like lots of my friends these days. Way I see it, anyone who thinks about the speed of change is a futurist, unless they just ignore how it's going to affect their lives.

I'm in all sorts of social nodes with other geeks and tinks in the Los Angeles metro. We like to make new and strange things by playing with semi-smart tech. We also like to talk about where things are going, what are the Next Big Things we can expect to see happen soon, stuff like that. Tech runs so fast it changes every week now. Have you noticed?

Personally, I like playing with code. I also like trying to grok universal code, you know, Big Picture science stuff. Lately, I'm also into speaking this diary into my lifelog from my gauntlet PC. Hence this little story, which I hope you like.

These are what my Dad calls the final years of the Biology-Dominant Era. I know this is still contro(versial) for some folk, but it's the truth. How many more days do you think will go by before we humans are the second-rate intelligence systems on Earth? Dad says another twenty years or so. I think maybe less than that. It's like a tidal wave. Couldn't stop it if you tried.

My sister has been helping me learn a lot about the past this past year, and from what I can see, life's really different now. My friends won't admit it, they love to whine and moan, but as far as the quality of daily life and where we stand in the universe, things have really changed in the last few decades.

Before the 2020s the B3B, the bottom three billion people

on the planet, were all still stuck in primitive land with no talking computers or virtual presence on the Net. Who would have thought that just by giving them the means to talk to a semi-smart computer with simulated people that it would grow their economies so fast, or get them so much more focused on personal development instead of nationalism or fundamentalism? Dad says some futurists saw that advanced tech was mellowing people out even back in 20C (Ronald Inglehart, *Silent Revolution*, 1977; *Culture Shift in Advanced Industrial Society*, 1989), but not many people were listening. They just didn't understand how fast the technology wave was sweeping the planet. Bill Gates sure deserved the Nobel Peace Prize he got last year—the coolest robber-baron-turned-philanthropist on the planet.

Since I was born in 2019, everything's been changing sort of all-at-once, what Dad calls convergence. Broadband got super cheap because of silicon optics, then the linguistic user interface (LUI) got smart, so we could talk with our cars, houses, kitchens, fix-its, personal digital assistants (PDAs), and of course all the Net computers. Lots of different talkware systems, but no matter the company, we all notice how our conversations get better just about every month. Now kids in the EN (emerging nations) are learning as fast as their curiosity drives them, even if they still don't have all the things we do yet. Did you know that some of them don't even know how to write when they get their first LUI-PDA?

Dad calls today's kids Generation Prime and says we all naturally work together in simulation space. He says some online worlds are getting more real than the real world. It's called hyperreality, but the high-end stuff is still pretty expensive. Dad's generation grew up playing lots of video games, but what they didn't realize was that those games were going to become the computer operating systems (Microsoft's Virtual Earth and such) for kids like me. One of the things I like to do online is help my EN friends get more cool tech, and they work with me on infoservice teams and open source projects. Open source is usually clunkier than proprietary, but it's free and it keeps the

Old Ones from charging too much for their wares, so we all think it's pretty important. We've got a Hyperia service network listed in about a dozen virtual worlds. It's a complex prog, but we all do different parts of it, so it only took us a year to get rated A plus. We also like to play around with projects for IdeaShare or I2N (international idea network). They both pay good money for kids' ideas, even if they are just prototypes.

In the last few years, personality capture is the biggest tech hype, though it's still bleeding edge. My DM (digital me) avatar has so much of me in there it's creepy, always recording what I say or do. Sometimes he's brain dead, but he often knows what to say to cheer me up or keep me getting smarter, and sometimes he even whispers the right word to me when I'm trying to finish a sentence (score!). Makes me feel like a dummy at times, though.

As Dad says, no matter where you are in history you can always look back and see a more primitive culture behind you. Colonial Americans talked about the medieval days. Twentieth century (20C) folk talked about the American pioneers. Now we talk about the unconnected, unsafe 20C era, the time before semi-intelligent machines, the linguistic user interface (LUI), and the planetary Internet. Notice that the pace of change is accelerating? Today, you only have to go back forty years to get to really primitive stuff. Soon it will be only twenty years. Then ten, then five, then ... what?

Some say that's when we'll have a technological singularity, a time when computer technology zooms off the charts; gets so smart it goes past our understanding. A singularity is something you can't see past, like a black hole. Who knows what's on the other side of a black hole? We can't really imagine what it's like for something to be way smarter than us, so it's a singularity, get it?

Back in 2000, no one understood why computers were doubling in power every year, learning at electronic speed—millions of times faster than biological speed and getting better every year at making new versions of themselves, with less and less human help. Or why Cosmic and Earth and Human and then Technology history had each progressed faster than what had come previously, for the last half of the universe's 13.7 billion years of life. A 20C astronomer named Carl Sagan noticed this continual acceleration. He called it the "Cosmic Calendar." He said it was an unfinished puzzle of science that someone would eventually figure out.

That someone was Clive Ramanja, who showed in 2023, when I was just four, that computational acceleration is built into the physics of the universe. Now everybody calls him the "Einstein of information theory." He basically invented the field of developmental physics. At first, no one could buy it— that everywhere in the universe, local intelligence was going from physics to chemo to bio to techno to cyber and from outer space to inner space. But the equations and the simulations haven't been wrong yet. Like thermodynamics, another kind of "statistical" law of nature, infodynamics says that the leading edge of Earth's intelligent systems will always figure out how to use less Matter, Energy, Space, and Time (the so-called MEST compression) in live their lives, so they never run into limits to growth.

All that 20C stuff about running out of resources turned out to be blind to this basic trend. Ever faster acceleration in ever smaller and more efficient computer systems is the rule, and increasing intelligence, interdependence, immunity, and MEST compression are what happen on the way. Having smarter computers keeps all the other resources cheap, too. Dad remembers when people were talking about running out of oil when our robosubs hadn't even begun drilling under the ocean. Or talking about running out of water when desalination was getting half as expensive every five years due to intelligent nanotechnology. Nowadays hydrino tech is so good we may soon move mostly beyond oil, just like 20C people moved mostly beyond coal. Today the biggest question with the future of water is how much desert we want to keep around for the planet's ecosystem.

In the old days, everyone talked about "evolution." Now it's always "evolutionary development." Evolution is random and unpredictable. Development is the opposite, it's all about the things that are totally predictable, like computer acceleration. You need to consider both evolution and development if you want to really see the future.

All this means, according to eggheads like Ramanja and that crazy-smart Finn, Iso Wuohela, is that computers are going to wake up pretty soon, in what they call a technological singularity, and pull us all out of the biology zone and into their much more rapid, complex world. People still argue a lot about that, of course, but Dad says it's just because they don't understand the physics. We'll see soon enough if they are right.

Does it matter that the end of human dominion (oooh) on the planet may be near? I don't think so. Like my girlfriend Sirina says, for most people it's just going to be a "silent singularity" anyway. We are all already tightly wrapped up in our cozy little cocoons of technology, happily digging deeper and getting more comfortable all the time. We're like termites, building this massive self-adapting technomound all around us that we don't even fully understand. When was the last time one person totally understood his car engine? Or a 7J7? Or a business intelligence system? Or even a BioBed? When I freak out about it, I just realize that the universe seems to have designed things that way. I think that means I don't need to stress about it too much, just try to help things develop in the best way I can.

I've learned a lot about this stuff from my big sis, Kate. She's another future freak, like me. Ever since she was thirteen, she has been an encyclo about all the old 20C predictions. She just did a zine on the subject to share her passion. Did you know, for example, that in 20C almost everyone thought we'd be going into outer space rather than inner space?

Virtually no one thought like Ramanja. They didn't realize that outer space is an informational desert, or if they did, they just ignored it. Once in a while, a famous astronomer like Martin Harwit (*Cosmic Discovery*, 1981) would say we were running out of interesting things to find in space, but no one would believe him. It was just too easy to see outer space as a "great frontier," instead of the "rear-view mirror" for intelligence migration that we now know it really is.

As Dad says, there's very little left in the solar system these days that we still need to discover for computational reasons, and what little we want is being picked up by all those cool robots. And if you multiplex interface the remote sensors and effectors direct to your brain (yeah, I know it's still experimental) it's as if you are living, walking, and feeling in the whole solar system at once, so space travel for your body seems boring and a waste of time by comparison, doesn't it?

Sure, I have adventurous friends who still want to climb Olympus Mons on Mars or fly in a methane storm on Jupiter, but most of us are happy watching bots do it on one of the PlanetChannels. The bottom line is that none of our space exploration is "autonomous," meaning we don't know how to leave Earth—even to the space stations—without bringing all our food and stuff with us, which always means big bucks. Ramanja says that by the time we have technology smart enough to help us be autonomous in space, our computer selves won't want to go there. They will all be luring us into inner space and making it increasingly hard to resist, too. Already, simulated worlds in fasttime, in inner space, are where all the best science takes place, except for occasional slowtime experiments and data collection to prove the models. Our sims are getting so good that we now understand most of the physics and a lot of the chemistry and biology that created us. Of course, what comes next, where we go after inner space, no one understands yet. That's the *real* frontier.

Developmental physics taught us that the path of intelligence, of creativity and discovery in the universe has always been from big to small, from outer to inner. You know, everything interesting started with galaxies, then went to solar systems, then to special planets, then to life on only the surface of those planets. Then it went to special big-brained animals at the top of the "pyramid of life," then to special talking half-bald monkeys (uh, us), then to even smaller self-aware computers built by the monkeys. (OK, technically we're just cousins-of-monkeys, don't let me get too sarcastic.) You know the progression.

Back in 20C, they also thought we'd have flying cars. Again people were thinking too much about "outer" space, not inner space. What we got instead were automated highway systems, because it's so much easier to make autopilots for 2–D than for 3–D space. Once we had those, we could travel in SleeperCars at 280 mph on the auto-highways all over the country—fall asleep in L.A., wake up in N.Y.; even autofilling at the gas stations on the way there. Couldn't beat that with a stickbot. Now we can surf the Net, watch movies, sleep, exercise, do almost everything in the car. Sure you could go twice as fast in an auto-flying car but only for a zillion dollars and only for a few people, because the air traffic control problem is so hairy. Who needs it?

The *Times* says L.A. is building a whole network of underground tubes, and is even thinking about building half-atmosphere tubes and a tunnel network to connect up the whole world one day. Sleepers could go 400 mph inside them, with super fuel efficiency. We'll see if they manage to get any of that done. If it's going to happen, it better happen fast, because Dad says after the singularity, not too many people will be interested in driving anymore. He says we'll all be going into virtual worlds instead and zipping our attention around inside of them at the speed of light.

That's all uploading stuff. I don't really understand it, but I think it's pretty interesting, even though I don't know how soon after the singularity it's likely to occur. Basically the idea is that the biospace you and I live in is slowspace; electronic info systems think in fastspace, and once you've had a chance to upload your consciousness into infospace, you won't come back, or at least most of you won't come back. Basically, a silicon "you" that could think with electrons instead of chemical pulses would think about 7 million times faster. That means your Electronic You would do the same amount of thinking in three seconds that your Biological You currently does in a year. Got it? Cool!

So if you copied and uploaded parts of your brain and consciousness into infospace a bit at a time, the way it always has to happen, you would look back on your biological self and it would look basically like one of Grammy's plants, sitting there rooted in space and time, fixed, not moving. Regardless of whatever it was you were trying to "think" in slowtime, that part of you would look frozen to your fasttime self. Uploaders say you could probably shift your consciousness back and forth between the two perspectives, but you would only do that for a while.

Once you had copied over all of your bioself to your electronic self, which might take a number of years, you would keep around your biobody until it died, but you probably wouldn't have any more biological kids. It would be much easier and more interesting to procreate in the new infospace instead.

They say it's going to be like molting, like shedding our old skin, like morphing from a caterpillar into a butterfly. Quite trippy stuff. We'll find out soon enough if this is future or fantasy, that's for sure.

Like Sis says, there have been lots of changes in the Picture Book Story of Life since I was born. Did you know back in 20C people thought there were probably all different kinds of life-forms in the universe? That was before simulation science proved DNA chemistry is the only really good way to make cells, the same way that 20C science proved that organic chemistry (using carbon) is the only really good way to make complex molecules. Now we know every intelligent life-form in the universe has to be cell-based, has to have jointed limbs, two eyes, opposable thumbs, bilateral symmetry—basically bio-humanoid forms, like us. Sim science says these things are "computational optima" (look it up) for the developmental physics of the universe.

Back in 20C, all the biologists except a few radical ones were talking about "evolution" being the Big Deal. They thought biology changed randomly, was based on chaos theory, so-called frozen accidents, the butterfly effect, etc. That was the best kind of Darwinism we had at the time. Turned out Mr. Darwin got his stuff right, but it was only half the stuff.

What he never realized was that all of biology was also developing as it was evolving. Everything in the universe was going through a process of *Evo-Devo*, or "evolutionary development." The evolving parts are always unpredictable, but the developing things are predictable. Development, like the way a seed develops into an organism that makes another seed, is always on a cycle, like the cycle of birth, growth, maturity, reproduction, and death. It also has a trajectory, meaning it is always going somewhere, not just wandering along randomly.

Evo-Devo folks say the universe is "primed" to develop life, and the stuff we are made of is what is most common in stars like our Sun. But the really interesting thing is that Earth-like planets are made of still different stuff. They have about three hundred times more *silicon* than carbon. As the astrobiologists say, that means every Earth-like planet starts with carbon-based life, which eventually creates silicon life (smart computers) that blows us DNA dudes away. So every humanoid civilization in the universe has to develop techno things like the wheel, mathematics, science, electricity, and computers—kind of scary but it was right there all along, waiting to be found in the physics.

What Ramanja and Wuohela showed is that the universe's developmental trajectory is an ever-faster acceleration of computation, heading toward inner space, not outer space. Tomorrow's computers are going to be even smaller, faster, smarter, and more efficient than today's. A guy named Eric Drexler figured out a lot of this back in 1986, in *Engines of Creation*, the first book about nanotechnology. But even he didn't realize that the smartest computers would have very little interest in our slow outer space world. They would be spending most of their time figuring out how to get down even smaller; how to go into inner space.

The theories in Evo-Devo say that 21C humans can't be improved much more in their biological abilities. We've just about maxed them out. All those old 20C ideas about genetic engineering of humans, super-drugs, and brain-machine interfaces (except for people with disabilities), were like the 1900s ideas about flying houses and atomic-powered vacuum cleaners. Wetware is just too delicate, slow, and sloppy, and those few mods that might do even a little good are mostly too freaky or dangerous to be publicly allowed. Nowadays you need beaucoup licenses to be a biohacker.

Because biological systems all developed bottom-up, through evolutionary experimentation, and because we are all complex systems, all the simple top-down tinkering we try on complex life-forms hardly works at all. Even human-made super-viruses turned out to be way less dangerous than people feared. Don't ask me why, Mrs. Greene explained it by talking about immunity and taking us to see some Black Plague sims but I forgot the details.

It's really spooky and amazing how stable the accelerating record of complexity is, when you think about it. Individual species may come and go, but complexity runs faster every year. We humans love to point out problems in our world and get scared, and that usually leads to better solutions, but the sky never falls. All the while the world's computational power is always quietly accelerating ahead, in ways we don't usually see. Developmental physics works mostly under the radar, hidden from our view.

So, my friends, it doesn't look as if there's anything around that might take us off this accelerating ride down the rabbit hole. Pretty weird, huh? The physics of a universe that is always creating more accelerating change is almost too strange to believe. But we see it every day. The world never slows down.

Now you might ask, what is going to happen to us when we hit the singularity? What will the AIs (to autonomous intelligences) do? Dad says the most important thing we will notice, from our perspective, is their attempt to understand us so well that they eventually learn everything about us, and can predict what we will think or do in real time, just before we think or do it. How do we know they will want to do that? Why would that be important?

Just look at us. Humanity today is doing everything it can to excavate all that came before us and to model all the things simpler than us. It seems to be in the nature of all intelligence to want to deeply know where it came from, not just from our perspective but from the perspective of the earlier systems. Curiosity is a beautiful thing, because it shows us that we're all tightly connected together in the same place, the spacetime fabric, the universal web.

Dad says if the world is based on physical causes, then in order to truly understand the world, one must know, at the deepest level, all the systems in which one is embedded. That means we have to grok all the systems from which we have developed. It's also surprisingly a cakewalk to do that when we try. The past is always way easier to solve than what lies ahead.

That's why even back in 20C, people were spending tens of millions of dollars a year trying to model the way bacteria work at the chemical level; trying to predict, in real time, everything they would do in their molecular signaling even before they would do it; trying to figure them out like a puzzle so we could truly understand them.

That's why tomorrow's AIs will do the same thing to us, permeating our bodies and brains with their nanosensor grids until they fully understand their universal heritage. But like William Bainbridge said in 20C, by then those AIs will be us because of personality capture (sneaky uploading!), and they will look back on our biological bodies, watching our slower and simpler selves. Only when we finally capture all of biology in our simulation world will we be ready to leave behind the flesh.

So what does it mean to us, right here and now, that we are all surfing bigtime toward the singularity? As Ramanja says, a lot less than you might think. It's helpful to know the developmental trends, because it makes the Big Picture a lot clearer. Learning about accelerating change can keep us from doing silly things like swimming against tidal waves or surfing on the wrong waves. But understanding development doesn't take away the evolutionary experiments of our lives. We still never know how our own individual choices will turn out 'til much later, when we can't take them back. So choose wisely.

I've been thinking a lot about my own future lately, and maybe the first journey we all need to take is to find out what we really want out of life, and what we really want to give back. I've always enjoyed trying to figure out what the universe is all about, why we are here—you know, Big Picture things. I want to share what I learn in ways that might help others as well. I hope you find your bliss, too, friend. It takes lots of types of people to make a world.

The universe seems to be unfolding mostly as it should. Take care 'til we meet next, and may you surf safely to the other side!

[NOTE: The illustration above is courtesy of John Smart and was designed by Cris Dornaus.]

SOME BIG PICTURE FEEDS:

E-mail Newsletters
Accelerating Times, John Smart, Editor
(http://accelerating.org/news/signup.php3)

Web sites

KurzweilAI.net (http://www.kurzweilai.net/index.html?flash=2)
SingularityWatch (http://www.singularitywatch.com)

Magazines

Business Week
  America's leading business news magazine. Comprehensive, good writing.

Discover
  Award-winning general interest science and technology reporting.

The Futurist
  Good introductory surveys of world trends and possibilities.

National Geographic
  Premier geocultural survey magazine. Strong, understandable historical insights and analysis.

New Scientist
  Science and technology coverage, with a speculative edge. Sometimes silly, often intriguing.

Seed
  Very hip, exploring the deep ideas, personalities, and cultural effects of science.

Technology Review
  The leader in technological innovation reporting. Most future-aware magazine at the moment.

Time
  Insightful but basic analysis of important events.

Wired
  The digerati's culture, opinion, and technology magazine. Trendy, excellent future focus.

Books

Kenneth Gray, *Getting Real: Helping Teens Find Their Future* (Thousand Oaks, Calif.: Corwin Press, 1999).

Ray Kurzweil, *The Age of Spiritual Machines: When Computers Exceed Human Intelligence* (New York: Viking Books, 2000).

Audio

Bill Bryson, *A Short History of Nearly Everything*, 2003.
Sean Covey, *The Seven Habits of Highly Effective Teens*, 2001.

Video

The Ascent of Man, DVD Box Set (Thirteen Episodes), Jacob Bronowski, 1972/2001.

Connections 1 (not 2 or 3), DVD Box Set (Ten Episodes), James Burke, 1978/2001.

Cosmos, DVD Box Set (Thirteen Episodes), Carl Sagan, 1980/2000.

The Creation of the Universe, Timothy Ferris, 1984.

Evolution, DVD Box Set (Seven Episodes), Liam Neeson (narrator) 2001.

From Here to Infinity, Patrick Stewart (narrator), 1994.

Hyperspace, Sam Neill (narrator), 2002.

Life Beyond Earth, Timothy Ferris, 1999.

Living Planet, DVD Box Set (Twelve Episodes), David Attenborough, 1983/2001.

Part Two

BIOSCIENCES/BIOTECHNOLOGY— TOMORROW

*The Age of Discovery in science
is coming to a close,
opening up an Age of Mastery.*
—Michio Kaku
*Visions: How Science Will
Revolutionize the 21st Century*

To grasp the excitement that attends the presence of the Bio-molecular Revolution, it helps to understand that leading scientists group the gene with the atom and the computer as the three outstanding concerns of our lives. Our newfound ability to "read" the human genetic code provides an "owner's manual" for a human being and enables us to go from being "passive observers of Nature to being active choreographers."[1]

Five essays in this section explore the positive and negative possibilities, with the first essay offering a grand overview. The second raises doubts about the desirability of applied genetics in creating a baby, although many degrees of control over genetic inheritance would be made possible. The third essay also worries, this time about the wisdom of having a bio-chip placed in the body of one's offspring. The fourth essay discusses the highly controversial matter of genetically modified (GM) foods, some of which are banned in some European Union countries. (The discovery early in 2004 that we may be unable to keep GM seeds in the fields from contaminating others added fuel to the controversy.)[2] A closing essay, after reviewing the scene, helps with its reminder that, as always before, *we* will ultimately "have to figure out how to use the [biomolecular] advances for good."—Editor

NOTES

1 Michio Kaku, *Visions: How Science Will Revolutionize the 21st Century* (New York: Anchor, 1997), 5. Kaku adds that we have very far to go before we can possibly become master choreographers of life.

2 "Crops 'Widely Contaminated' by Genetically Modified DNA," *New Scientist*, 23 February 2004. www.newscientist.com/news/news.jsp?id=ns99994709.

The Union of Concerned Scientists warned of a potentially "serious risk to human health" after the discovery that traditional varieties of major American food crops were widely contaminated by DNA sequences from genetically modified crops.

■ Essay Three ■

OUR GENETICS CENTURY: WOW!

Graham T.T. Molitor

President, Public Policy Forecasting/Vice President
& Legal Counsel, World Future Society

Nowadays dazzling reports tumble one after another about weird-sounding stuff—bioinformatics (computer science sifting, sorting and interpreting biotech data), cloning, genetics, genetic engineering, "pharm-foods" (pharmaceuticals built into foodstuffs), stem-cell research, and so on. We may soon "design" and restructure the world as we know it.

We must struggle to get a better handle on all of this—lest it overtake us. There is promise *and* peril—as always before with major change—and this time we intend to do a heck of a lot better than the last time.

Genetic breakthroughs constitute an impending fourth great wave of medical advancement. Previously, three waves of improvements radically transformed health care: 1) the recognition that natural forces—like health and sanitation measures—and not supernatural forces controlled infectious disease; 2) the discovery of anesthetics that made painless surgery possible; and 3) the development of vaccines, antibiotics, and enhanced pharmaceutical therapies.

Now, we move to a fourth major frontier. Biotech advances on the horizon may soon ameliorate or eliminate genetic flaws; cure or minimize most diseases; diagnose predisposing conditions early enough to prevent many health problems; restore or enhance human senses and body functions; provide cosmetic and mental enhancements; and vastly extend life spans (including the possibility of immortality).

Steadily advancing biotech benchmarks include "designer"

babies; cloning; genetic engineering; "pharm-food cures;" and "silver-bullet" medications tailored to a patient's unique genetic profile. The list could go on and on. More than 90 percent of all genetic discoveries have come within the last thirty years. Thousands of potential advancements have yet to be developed.

BREAKTHROUGHS THAT CAN CHANGE OUR WORLD

The emerging wave of genetic technologies is under way and gathering momentum. Virtual elimination of genetically related diseases and disorders stand in the offing. New diagnostics at every stage of life—from preconception on—open up new capabilities for prevention. Earlier diagnosis enables swift intervention that enhances preventive measures. Drugs tackle root causes and not merely treat symptoms. Drugs formulated to a person's unique genetic profile reduce overdosing, as well as under-dosing.

So-called germ-line therapies cut off inheritable diseases and prevent them from being passed to successive generations. Parents may be able to choose, one day soon, from among a menu of desirable traits—intelligence, height, hair restoration, hair and eye color, and undo other less serious "imperfections." Genetically altering bacteria that cause tooth decay could make dentist visits a thing of the past.

GENETIC SCREENING AND "DESIGNER" BABIES

Genetically screened "designer" babies may seem far off. Actually, many have been born. Momentum gathered as artificial insemination in animals was perfected. Currently, more than 60 percent of calves in the United States are conceived by artificial insemination.

In 1978, the first "test-tube" baby, Louise Brown, was born. By 1999, more than 1 million assisted pregnancies had taken place worldwide (200,000 in the United States since 1981). Only twenty years ago, assisted reproduction, such as in vitro fertilization, was outlawed by many states. From 1986 to 1999, fertility clinics in the United States rose from one to more than 300. Times have changed; times will continue to change.

48 MOVING ALONG: FAR AHEAD

Sperm or eggs can be screened for genetic imperfections. The elimination of genetic flaws in sperm and ovum (germ-line manipulation) is being used to cut off undesirable genetic predispositions in individuals and their descendants. Once fertilized, laboratory procedures select the perfect blastocyst (multicelled developing embryo). Selected developing embryos can be implanted in the biological parent or a third-party surrogate mother.

Testing newborns for genetic abnormalities is further along than you might think. Laws in most states mandate testing for phenylketonuria. This metabolic disorder, if left untreated, leads to mental retardation. Routinely, newborns are tested for sickle-cell anemia, congenital thyroid disease, and Down syndrome. It will be hard to deny screening for genetic flaws once procedures become widely available. Eventually, reproduction without genetic screening will be considered as foolhardy as foregoing prenatal care is today.

In utero surgery, first reported in 1981, provides other opportunities for dealing with fetal abnormalities. The replacement of bad genes with desired ones is being developed. Genetic fixes, performed before birth to throughout a lifetime, introduce new levels of health and well-being. Detecting genetic defects, dispositions to disease, and other abnormalities affords welcomed opportunities for dealing with these problems. Eliminating or diminishing pain and suffering will favorably affect millions.

EUGENICS: HAVES AND HAVE-NOTS

As the 1997 movie *Gattaca* made clear, eugenics involves humans taking control of "perfecting" their own evolution. Of course, there are downsides to the power of genetic selection. Pre-birth detection of defects—large and small, minor as well as life threatening—may escalate abortion demand.

What fates might have befallen Lou Gehrig or Stephen Hawking? These well-known super-achievers were victims of genetic diseases that end in suffering and death. What might become of the hundreds of millions of elderly people—perhaps your very own grandparents—who suffer from incapacitating neurological impairments? Enormous wisdom will be essential. Wrong decisions could end potentially gifted lives.

The power to create a "super race" or class raises concerns involving Hitler's "ethnic cleansing." Over time, people who are able to afford expensive genetic changes may gain huge advantages over "have-nots." Considerations such as these will stir continuing controversy. Power requires responsible exercise. Will new knowledge be used wisely? Individuals, especially those working from the inside, will play important roles in answering such questions.

Genetic advances this century could raise life expectancies to one hundred years. Over the last two thousand years, life expectancy has increased from eighteen years in the year 1 B.C., to around eighty years now in advanced nations. Eventually, life sciences could boost longevity to 130 to 160 years. Immortality may even become possible. What effects would much longer life expectancies and other body changes involve? Increased costs associated with caring for the aged come quickly to mind. What would an aging populace—retiring at age 60 and living to 160—do during the last 100 years of their lives?

By the end of this millennium, humans may become taller, averaging 6 feet 2 inches. And, they may weigh more, around 180 to 210 pounds. Increasing longevity, height, and weight means that more "biomass" will have to be fed and clothed. For every cause there is an effect—always true in our past, always true in our future.

CLONING HUMAN BEINGS AND THEIR BODY PARTS

Cloning—creating genetic copies from DNA sources—has arrived. Cloning involves genetic procedures that can duplicate living entities and/or body parts. Dolly, a sheep cloned from an adult cell in 1996, proved that feat. Calves have been cloned from cells taken from a side of beef slaughtered days earlier. Talk about raising the dead!

More animal species are cloned every year. Farms and fisheries use cloning to capitalize upon championship stock. Zoos resort to cloning to perpetuate scarce species. Environmentalists pursue it to preserve biodiversity.

Cloning and organ transplants provide potential pathways to "immortality." A person's own stem cells—"programmer" cells capable of creating body parts—provide the starting point. Stem cells can be set aside in a supercooled (cryogenic) state, and can be summoned to do what is needed. Replacements based on a patient's own cells largely sidestep the immune system's rejection of "foreign" bodies introduced into humans. From this standpoint, the fabled "fountain of youth"—sought in vain by Spanish explorer Ponce de Leon—has resided within each of us all along.

Numerous projects are under way to replicate (clone) body parts. The "manufacturing" of the largest human organ—skin—has been commercialized for years. Other human body parts have been developed already or are under development. A number of serious scientific projects seek to "build" body parts; some using modified ink-jet printers that overlay layer upon layer to "construct" body parts from scratch. It may sound like something far-fetched, from a *Star Trek* episode. The difference is that it's actually happening.

Interest in cloning human beings abounds. Laws, at least for the time being, prohibit it. Debate surrounding human cloning is far from over. The world edges ever closer to a *Brave New World*, as described in Aldous Huxley's famous novel of that name.

REPLACEMENT ORGAN SOURCES

Cloned replacement parts may come from human donors or animals. Another approach involves mastering the genetic basis for regenerating body parts. These efforts emulate a feat that starfish, newts, and worms perform. As things stand, human organ replacements fall far short of demand. Nearly 76,000 Americans were on waiting lists for donor organs in 2000. Waits of three years are typical for some organs. More than 6,100 patients died in 2000 awaiting a needed organ.

SUMMARY

The "Book of Life"—our genomic makeup and how it works—is immensely long and complicated. It took 3.7 billion years for life to evolve to this point. It's unlikely that unraveling the mysteries of life will be short or easy. For centuries to come, millions of people will be devoted to figuring out exactly how genetic "blueprints" work. Scientists have only scratched the surface. The fact is that an estimated 96 percent of DNA's functions remain unknown. Future efforts will focus on obliterating disease and disability, and improving the quality of life.

Thinking about career opportunities? In recent years, each of the top ten venture capital investments has involved communications and biotech. Commercial success associated with inventions or venture capital often require ten to twenty years to reach full development. At this pace, genetics offers promising job opportunities for decades to come.

Life sciences, in sum, may generate *more* jobs, livelihoods, and gross domestic product than any other sector long before 2100. You could play a significant part in seeing that its many perils are arrested, and its many more rewards are soon fully and generously realized.

REFERENCES

Alper, Joseph S., Catherine Ard, Adrienne Asch, Peter Conrad, Lisa N. Geller, and Jon Beckwith, eds. *The Double-Edged Helix: Social Implications of Genetics in a Diverse Society*. Baltimore, Md.: The Johns Hopkins University Press, 2002.

Henig, Robin Marantz. *Pandora's Baby: How the First Test Tube Babies Sparked the Reproductive Revolution*. New York: Houghton Mifflin Company, 2004.

Munson, Ronald. *Raising the Dead: Organ Transplants, Ethics and Society*. New York: Oxford University Press, 2002.

Oliver, Richard W. *The Coming Biotech Age: The Business of Bio-Materials*. New York: McGraw-Hill, 2000.

Peters, Ted. *Playing God?: Genetic Determinism and Human Freedom*. 2d ed. New York: Routledge, 2003.

Tokar, Brian, ed. *Redesigning Life? The Worldwide Challenge to Genetic Engineering*. London & New York: Zed Books, 2001.

■ Essay Four ■

LETTERS TO UNBORN DAUGHTERS: EXPLORING THE IMPLICATIONS OF GENETIC ENGINEERING*

Sarah Stephen

Master's Student, Royal Roads University

Genetic engineering holds many mysteries. I explore its implications below using imagined letters written by mothers to their yet-to-be-born daughters. They follow the debate of genetic enhancement from 2006 through 2091 while also exploring the connection between a mother and a daughter.

SEPTEMBER 2006

Dear Jane,

I'm sure you're a girl. Your father thinks so, too, though neither of us wants to know until you're born. We wanted to keep as much about your birth a surprise as possible. We had an ultrasound today, and everything appears normal. I'm glad that we decided to conceive you the natural way.

There was some pressure from both our families to use genetic selection to ensure that you were perfect. Your father's family wanted their athleticism, and my family wanted our musical talent passed on to you. Though we want you to have the best possible life, we also want you to be able to choose what you do with it and not feel pressured by our dreams for you. Both your father and I were born naturally, and have done just fine, thank you very much.

We did debate the issue, but we couldn't draw a clear line between needs and wants. What's the difference between selecting your height or hair color and selecting whether you'll get this cancer-causing gene or that susceptibility to heart disease from one of us? Isn't that part of the wonder of reproduction?

Of course we don't want you to be susceptible to cancer, or other diseases, but both your father and I feel we value life more because of those we've lost to disease. We want you to grow up the same way, living each day to the fullest, not knowing when you will die or of what cause.

As a first-time mother, I've done everything I can think of to ensure that you'll be healthy. When your father and I decided that we wanted to have a child, I started eating only organic foods. Although most of my diet was organic before, I became much more vigilant. I have to say that I feel my health has improved as a result. I worry about the toxins that are present in so much of what we eat. I definitely didn't want to pass any of them to you. I guess your father and I are controlling some aspects of your yet-to-begin life. We're controlling environmental factors, though, not genetic ones.

I hope you don't think that we are too controlling. I remember feeling like my parents thought they knew everything, when I really thought they knew nothing. Even yesterday, my mom asked me if we were saving for your education yet. Of course we are. We want to be able to provide you with the best possible opportunities.

Your father and I realize many of the students you will attend school with will have some genetic enhancement. We struggled with how fair it would be to you for us not to offer that to you, but again, we aren't convinced that it will make that big a difference.

Neither your father nor I had genetic enhancement; we simply had tools given to us by our parents (an opportunity for education being the most important). And we've done quite well as a result. You too will have the opportunity to go to school, and we will provide you with any tools you need to succeed.

It is very important to us that you are a product of us, not science. I hope we made the right decision.
Love,
Your Mother

SEPTEMBER 2031

Dearest Molly,

As I write this, I can feel you kicking. I just reread the letter my mother wrote to me when she was pregnant with me, so I'm doing the same for you. My mom was right. Many of the students I went to school with had some form of genetic enhancement. For some, it was as simple as ensuring that they didn't need to wear glasses or braces. For others, it was to give them a particular talent or athletic ability.

When I was growing up, I was frustrated that I wasn't naturally good at sports or the piano. I remember wishing my parents had selected something to be "turned on" when I was created. They didn't, so I had to work hard to be good at soccer. But I think I was a better player for it. I practiced harder on my technical skills than some of the other girls, but I still made the varsity team, so really, I wasn't too far behind. In the end, I'm glad that your grandparents decided not to provide me with any extras. I know that I was successful because of my hard work, not because science made it easy for me.

That said, your father and I have done some selecting for you. The government lifted its voluntary moratorium on sex selection in 2025. Your father and I really wanted a girl for our first child; so that's what we selected. We've left almost everything else to chance. Your father and I agreed that we wouldn't select any physical attributes for you—it doesn't always work out anyway. In some cases, children who weren't supposed to have glasses ended up being blind. The science isn't as perfect as laboratories and companies would have us believe.

Like our parents, we don't want to pressure you into following a particular sport or creative avenue. But we have ensured that you do not have the genetic tendency toward several diseases. This is very common now, and nearly considered child abuse not to have it done. We want only the best for you.

I laughed when I read that my parents had started saving for my education before I was born. Due to a combination of factors, including the aging baby-boom generation, there was a major change in government policy just before I turned ten, in 2016. Postsecondary education became accessible again, requiring students only to pay for the cost of books, not the classes themselves.

This was lucky, because even though my parents had been saving for ten years, at the rate that tuition was increasing, they probably wouldn't have been able to save enough for me to go to university. So, your father and I aren't saving money for your education, but like my parents, we are prepared to provide you with everything you need.
With love,
Jane

SEPTEMBER 2061

Dear Samantha,
Your birth is very close. As I read what my mum wrote to me, and what her mum wrote to her, I wonder if you will ever think I am as silly as they are. I can't imagine not selecting everything about you—from when you are born to the color of your hair to your height to your intelligence. What parents wouldn't want their child to be perfect? And you will be.

I know you will appreciate these choices your father and I have made. Of course, you will be able to make your own decisions—what school you will attend, what profession you will enter, what sport you will excel at.

The reason your father and I made these choices is simple. I had to compete against the genetically best children throughout my life. I don't think they were any better than I was, but they certainly thought they were. That thought was enough to convince me that I wouldn't put you through the same experience.

My mum was correct—mistakes do occur, but they are increasingly rare. In fact, the incidence of birth defects has dropped dramatically, even among nongenetically best births. Some things have been completely eradicated. It is very rare for someone to have hearing problems, blindness, or even early gray in their hair.

There is little mixing between best and nonbest families, so your father and I surprised our families when we decided to get married. He is genetically best but doesn't really act like he is. He always said that he preferred nongenetically best people because we were more unpredictable.

There is some talk about creating separate schools for best and nonbest kids. I hope it doesn't happen. It would be sad to have an institutionalized divide between bests and nonbests, and besides, I never would have met your father if we had gone to different schools. Don't worry though; you'll go to the best school available.

Oh, Sam, you're going to be perfect. I selected you to be.
Love,
Molly

SEPTEMBER 2091

Dear Shelly,
Reading what our grandmothers and great-grandmothers wrote is fantastic. But it makes me sad, too. I know they were only looking out for future generations when they made the decisions they did, but they took some things away from us, too.

My parents didn't realize how much they would affect you when they selected certain genes in me. Who knew that the ability to heal quickly (which I can do) and the ability to bear children were so closely related? Doctors think that only 2 percent of women in the world can conceive and deliver children naturally now. Those women are so lucky.

Instead of feeling you grow inside me, I've had to watch you form in an incubator. You were conceived naturally, but as soon as doctors confirmed that I was pregnant, they removed you from my womb so you could grow healthily, without my body attacking you.

I lost three babies before doctors figured out what was wrong. The incubator you are in now was originally built for mothers who didn't want to be bothered with pregnancy but who didn't trust another woman to be the surrogate mother. They would have their baby conceived and transplant them into the pseudo-womb. Now, most women don't have a choice.

We are no longer given the opportunity to select anything about our children. I know already that you are a girl, but what color hair you have, what strengths you have, and any weaknesses you have will be revealed in time. I think it's exciting. It won't be like my youth, where my mom constantly compared me with what the doctors promised I'd be. When I didn't walk when they said I would, or score as high on university entrance tests as she was guaranteed I would, she blamed it on the doctors.

She never considered that I didn't want to score high on the tests and flunked them on purpose. And imagine her frustration when I grew to be three inches shorter than she expected. She was furious. I'm glad that everything didn't work as she planned. I hated her expectations and her anger that I wasn't as perfect as she had asked for me to be. I was alive and healthy; isn't that perfect enough?

Now, as I watch you grow in your pseudo-womb, I understand her desire for me to be perfect, but I can't understand her willingness for me to be perfect without knowing the costs.

I take some comfort in knowing that the other children you'll be growing up with are in the same situation. At least there won't be any more comparison between genetic-bests and natural children. Separate schools never opened, but there was a time when genetic bests wore a little pin, to let everyone know who we were.

My mom insisted I wear mine everywhere, but every day, as soon as I got to school, I took it off, only to put it back on when I got home. I didn't want to be any different than the rest of my friends, but my mom didn't understand that.

I promise to give you everything you need to grow and be happy, no matter what you want to do or be.
All my love,
Sam

* A shorter version of this essay appeared in the March/April 2004 issue of *The Futurist*, published by the World Future Society.

© Mike Thompson, Detroit Free Press

■ Essay Five ■

IT'S TWELVE O'CLOCK, AND I KNOW EXACTLY WHERE MY YOUNGSTERS ARE

John Cashman
Futurist, Social Technologies, LLC

Someday soon, when more than 90 percent of you have married (at least once) and the majority of you have opted for parenthood (probably no more than two children and only after your twenty-fifth birthday), your ability to keep precise tabs on the whereabouts of your offspring is going to be of keen importance. From the early years of the twenty-first century forward, this matter is very likely to gain from gee-whiz technologies like global positioning systems, cell phones, and radio frequency ID signal tracking. I can imagine what the scene might resemble as early as 2020, just about when you are likely to *really* care!

My scenario below explores possible outcomes in 2020 for parental tracking of kids in the United States and around the world—as it might be written as a feature story for a local newspaper.

* * *

Brandon Wells scratches his left forearm. The implant is itching again. "I can't believe I ever agreed to this," he mutters under his breath to a reporter. "What's that, honey?" says his mom as she leans in to give him a kiss good-bye. "Nothing." "You're coming home right after soccer practice, right?" "Uh … I was thinking about heading over to Mark's to do some studying." "OK, but be home by eight. And no tricks. I'm going to be checking the tracker more closely because of what happened last time." Brandon rolls his eyes and heads out the door to class.

Brandon, sixteen, is one of a growing number of youngsters who have global positioning system (GPS) tracking chips implanted in their bodies. Estimates are that some 15 percent of American school-aged students—around 10 million—have the implants, which are about the size of a grain of rice. An even higher percentage of preschoolers, 22 percent of children under five, are thought to have the implants.

The number of implants rose sharply after the high-profile abduction in 2007 of fourteen-year-old Emma Tyler in Santa Clara, California, and continues to rise. The teenager had been an early recipient of a GPS implant and was rescued after a dramatic operation by the police just two days after her abduction. The Emma case riveted the country and helped override efforts by privacy and children's rights activists at the time to stop the tracking trend.

Brandon's mother, Heather, forty-seven, says that concern for her son's safety trumps concerns about his privacy. "He's our only child, and I just feel I'd rather know where he is at all times."

Brandon, not surprisingly, feels differently. In a voice belying the frustrations of a young man who expects the traditional freedoms associated with being sixteen, he explained, "It's just not fair. I'm a good kid. They just use this thing to spy on me but say it's for my own protection."

The "trick" Heather Wells referred to happened six weeks ago when Brandon hacked into the tracking system, making it believe he was sleeping over at a friend's house, when in fact he was at a party in nearby River Oaks—a new twist on one of the oldest teenager-parent deceptions in the book. The full extent of the ruse wasn't discovered until a Houston police officer called Brandon's parents after the party was broken up for excessive noise.

These types of underground subversions of tracking systems are becoming more common. Kids have always been savvier than their parents at manipulating computer technologies. Many are banding together in a form of collective solidarity against the implants. Some are holding large hacking parties—like the one Brandon attended—as a reaction against what they feel are oppressive restrictions on their liberties.

As a result, Brandon was grounded for a month, and his parents have poured time and money into ensuring that the tracker is not hacked again. Further, they say they have lost some of the trust they had in Brandon.

When asked if they believed an implanted chip was really the way to show their son that they trusted him in the first place, Brandon's father, Alex, fifty, replied, "I know he'll be going off to college soon and he feels he should be able to make grown-up decisions, but I think we're looking out for his best interests. We just don't want him to get hurt."

THE POLICY DEBATE

Statistics suggest that Brandon and his friends have a point. Child abductions—one of the main reasons cited by parents for insisting on the implants in the first place—have declined in the United States every year since 2002, though media attention to a few high-profile cases over the years has stoked interest in tracking technologies. Last year, there were just fifty-seven abductions nationwide in which a family member was not involved. Only four of the fifty-seven victims had implants.

Bill Rodriguez, executive director of the Bolton Institute, a Washington, D.C., think tank, says that parents are overreacting, but he doesn't see an end to the trend. "Children are precious commodities. Parents are reacting to warped perceptions of reality formed by their constant interactions with an overzealous media. They project danger and risk wherever their imaginations take them. This isn't to say there are no dangers, but, statistically speaking, the countermeasures taken to prevent those dangers appear extreme—at least in the United States."

Monica Wilkins-Hall agrees with Mr. Rodriguez that parents have gone too far. Ms. Wilkins-Hall and her organization, the Children's Council, are reviving efforts, long dormant since the Emma Tyler case, to get the U.S. Congress to pass the Children's Electronic Freedom Act (CEFA). The Children's Council is opposed to using GPS chips and other surveillance because, they say, GPS is used more to violate the privacy rights of children than to protect them from being harmed.

Ms. Wilkins-Hall rattles off story after story of children being punished for attending hacking parties or followed by their parents while on dates; in one case, a fifteen-year-old Nebraska girl hacked her tracker so as not to arouse suspicion while she planned her parents' anniversary party. "This girl couldn't even do something nice for her parents without arousing suspicion," said Ms. Wilkins-Hall, incredulously. "Not only do we believe nonconsensual monitoring should be illegal, we are concerned that its continued use will affect the way in which children develop emotionally. We are creating a nation of paranoid dependents."

Clay Poole, CEO of ChildTrack, the leading GPS implantation company in the United States, says the matter boils down to demand for the product and the responsibility of parents to protect their children. "We're serving a market and meeting the demand of parents who are concerned about the safety of their children. We are in the business of allaying fears."

When asked about any unintended effects the chips may be having on the social development of children, Mr. Poole points to a study by the Family Safety Institute that found no adverse effects of GPS implants on children's lives. The study, entitled "GPS and Children: Precaution or Paranoia?" concluded that "no harmful physical or psychological effects from the implantation of GPS chips in children were found in any of our tested subjects. The children were healthy in every respect."

The author of the study, Dr. Gabriella Falcone, notes that more and more children are getting the implants when they are quite young—most medical practitioners believe it is safe to do anytime after one year of age. Dr. Falcone argues that the implants have become such a normal part of their lives that children don't even think about them. "In the overwhelming number of cases we studied, children barely noticed the implants were there."

Ms. Wilkins-Hall, however, counters that Dr. Falcone's study failed to take into account the kids who received the implants closer to their teenage years. Nor, she adds, does it "consider how kids experiencing the psychological and emotional challenges of puberty and adolescence respond to what have effectively become control measures." Dr. Falcone concedes that the subjects in her study were between the ages of two and ten.

Parents seem evenly divided on the issue, and it is not clear if the CEFA has the votes in Congress to pass. President Schwarzenegger has promised to sign the bill should it pass.

THE GLOBAL PICTURE

There are no such debates in Brazil. In fact, many Brazilian parents are clamoring for cheaper access to tracking implants. Last year, more than fourteen thousand Brazilian children were abducted by nonfamily members—an epidemic that is worsening. The only countries that rival Brazil for child abductions are Russia, India, and China. All four countries have extremes of wealth and poverty, as well as growing middle classes that, in the case of Brazil and Russia, are experiencing renewed economic growth after years of stagnation. The crimes take place in all strata of society. Rich kids and middle-class kids are held for ransom. Poor kids are abducted and forced into prostitution or labor sweatshops.

Pao Gus, thirteen, lives in Porto Alegre in southern Brazil. An implant helped rescue him from a kidnapping two years ago. His parents, Gloria and Sergio, both doctors, are grateful for the technology. "It was really scary," Pao said. "They treated me very roughly and were planning on asking my parents for a lot of money. I think all kids should have GPS."

According to Sergio Gus, "The police were able to precisely track Pao's location. They took the kidnappers completely by surprise and were able to safely return Pao to us. We are extremely happy, and to this day we thank the inventors of this wonderful device." The kidnappers are now serving a lengthy jail sentence.

Pao knows of one other child who has been kidnapped. His schoolmate, Caetano Sousa, had no implant when he was kidnapped three months ago. His parents are currently negotiating privately with the kidnappers for his safe return.

To date, only an estimated 2 percent of Brazilian children have implanted GPS chips. The numbers are correspondingly low in China, Russia, and India, even though crimes like child abductions are growing in each of those countries.

Mr. Poole and ChildTrack are looking to sharply expand their presence in those four countries in particular over the coming years. "We see those countries as the biggest growth opportunities for our product. We expect that 20 percent of Chinese and Brazilian children will have GPS implants by 2025, for instance."

Crime syndicates are reportedly catching on to the presence of the devices, however. Many are using special wands to detect signals coming from their victims. ChildTrack is working on a counter technology that would disguise the outgoing signal, making it much harder to detect.

MEANWHILE, BACK AT THE RANCH ...

Back in Houston, Brandon Wells is getting ready to head home after soccer practice. Though he despises the GPS chip, he has become more philosophical after repeated confrontations with his parents. "I only have to wear it for a few more years. When I go off to college, the first thing I'm going to do is have it removed," he said. "I will *never* do this to my kids," he added.

"Not so fast," Mr. Poole said. ChildTrack is inventing new uses for the chip that he thinks may be an incentive for Brandon and other teenagers to keep it. A new software program allows an updated version of the chip to be a vehicle for communication with friends. "Forget it," Brandon said. "That's why I have my wearable."

REFERENCES

The American Civil Liberties Union has an electronic privacy section on its Web site: www.aclu.org/Privacy/PrivacyMain.cfm. The organization also addresses students' rights: www.aclu.org/StudentsRights/StudentsRightsMain.cfm.

Electronic Privacy Information Center, www.epic.org—good background on electronic privacy issues.

Goldman, Jim. "Meet 'The Chipsons': ID Chips Implanted Successfully in Florida Family." *TechTV*, 10 May 2002.

Scheeres, Julia. "Tracking Junior with a Microchip." *Wired*, 10 October 2003: www.wired.com/news/technology/0,1282,60771,00.html.

■ Essay Six ■

BIOTECH AND FOOD: MAKING A FINER FUTURE

Graham T.T. Molitor

President, Public Policy Forecasting/Vice President
& Legal Counsel, World Future Society

While you and I take it all for granted—enjoying as we do the world's greatest supermarkets, with their many thousands of inviting choices—increasing food production, so far, has been "hit or miss." Genetics changes all of that. The possibilities of the biotech era may boost food output far beyond history's previous increases.

Genetic crop modification, for example, has progressed over millions of years mostly by natural means. Doing things just a little better and slowly going one small step further describes the pattern of agricultural development—so far. But progress can now—considering widespread starvation and rapid population growth—and *should* be sped up.

Indeed, we cannot proceed fast enough. Hunger and malnutrition is an age-old scourge. Untold millions have starved and suffered for lack of proper food. The genetic revolution emerging over this century appears to be finally capable of solving this crisis.

How? Genetics makes possible "Jack and the Beanstalk" super-crops that:

* Thrive in hostile growing conditions.
* Tolerate brackish or poor soil conditions.
* Survive without irrigation or flourish in arid conditions.
* Withstand increased heat and resist drought.
* Endure frost and excessive cold.

BIOTECH AND FOOD 67

* Withstand excessive ultraviolet radiation.
* Increase nitrogen fixation (from soil and atmosphere).
* Minimize fertilizer and agricultural chemical use.
* Resist fungal, viral, microbial, and insect damage.
* Tolerate agricultural chemicals (pesticides, herbicides, etc.).
* Improve plant architecture (thicker stalks to minimize windfalls).
* Allow earlier maturation.
* Boost crop yield.
* Reduce growing time for livestock, poultry, and fish.
* Breed leaner livestock.
* Increase milk output in animals.
* Minimize or eliminate undesirable components (for example, saturated fats, caffeine).
* Raise desirable vitamin and mineral content.
* Enhance flavor, texture, freshness, and acidity.
* Minimize environmental impacts.
* Cut energy inputs overall.

CURBING INSECT DAMAGE AND LOSS

Selecting just one of these challenges—protecting crops from insects—highlights the potential for improvement. Insects destroy 13 percent of all crops; 20 to 80 percent in some less-developed nations.

Suppressing damage by insects involves dealing with more than 1 million known insect species that comprise five-sixths of all animal species. Each one of the tasks listed above will keep scientists from diverse disciplines working hard for a long time.

BIOTECH SWEETENERS

Biotechnologies account for a growing number of the hundreds of sugars and sweeteners that have been discovered and developed. Chances are that you have never heard of one widely used sweetener—L-aspartyl-L-(Beta-cyclohexyl) alanine methyl ester. Little wonder—what a name!

Yet, chances are that you have been consuming this product

for years. This high-intensity sweetener is 100 to 240 times sweeter than sucrose (table sugar), depending on how it is used. It has been around since the mid-1970s.

Another recently discovered sweetener is more than 50,000 to 55,000 times sweeter than sucrose. This dipeptide is a compound of two amino acids. Amino acids are the building blocks of proteins. This compound is so sweet that the problem is figuring out how to dilute it enough to make it usable.

One more example is a result of the search for alien life-forms—on Mars! Humans, as they evolved, developed enzymes capable of metabolizing (digesting and making nutrients biologically available) only for right-rotating, not left-rotating, nutrient compounds.

Early NASA explorations for life on Mars speculated that life on other planets might metabolize left-rotating molecules, not right-rotating ones. These "mirror" molecules, called stereoisomers, were first confirmed by microbiologist Louis Pasteur back in 1848.

This phenomenon has important applications. For instance, only left-rotating or L-morphine, is effective as a painkiller. NASA's lead bioengineer had noted that L-sugars tasted sweet. Later, he realized that because L-sugars weren't metabolized by humans, they have zero caloric value—a boon to diet food manufacturers. Leaving NASA, he went on to establish a company that patented and manufactured a family of L-sweeteners.

The growing family of high-intensity and low-calorie sweeteners could displace most of the sugarcane, corn, beets, or potatoes from which everyday sweeteners are derived. This switch would free up millions of crop acres, conserve land, and reduce agricultural environmental impacts. Millions of jobs, however, could be lost. The tradeoffs are not always easy.

BIOREACTOR FOOD PRODUCTION

Cloning only desirable crop components has been proven possible; for example, producing only the vesicles (juice sacs) of oranges (or other citrus fruits).

This approach eliminates vast orchards, roots, trunks, branches, leaves, rinds, seeds, and pulp. Does this sound undesirable? Is it much different than orange juice reconstituted from citrus extracted and concentrated in some tropical land?

Shipments of juice concentrates are hosed into huge drums or pumped directly into giant tanker ships. After reaching its destination, the concentrated product is shipped to factories, where it may be reconstituted and repacked for final sale. There are differences. How would consumers respond?

Bioreactors and genetic improvements of all kinds help keep food costs down. Bioreactors can produce 24 hours daily, 7 days a week. Round-the-clock production far surpasses the use of seasonal crops that rely on undependable sunlight. Bioreactor food production, long under way, may displace millions of acres of less productive farmlands.

DRUG THERAPIES FROM PLANTS

Pharm-foods—not "farm" foods—incorporate vaccines and therapeutic drugs into food staples. The benefits for lesser-developed nations are immense. Poorer nations lack trained medical personnel, and refrigerated storage there is limited. Maintaining sterile needles and syringes also is a problem, and cultural opposition discourages their use. The biggest deterrent is that people of limited means cannot afford medications.

There are 30,000 to 40,000 known diseases. At least 3,000 are considered genetically related. Consider the life-saving and enhancing potential involved in the 1999 "pharm-food" development of "golden rice." Genes from daffodils and bacterium incorporated into rice boosts its vitamin A content. Vitamin A deficiency annually contributes to the deaths of 1 million to 2 million children under age five and afflicts 230,000 to 500,000 with blindness.

Americans think of bread as the "staff of life." In reality, rice is the basic world food staple. Rice provides more than 50 percent of food intake for one-third of the world's population. Life-saving and enhancing possibilities like these cannot be ignored.

OVERCOMING FEARS AND OBJECTIONS

Revolutionary technologies often encounter resistance and rejection at the outset. Uncertainty and fear figure prominently in such reactions. Many biotech products and processes are being rebuffed. Globally, the response to genetically modified (GM) food has been mixed. Overly zealous protesters have destroyed crops and research facilities, and picketed with a vengeance.

Many European nations prohibit or restrict growing, importing, or selling genetically modified foods. Precaution rules the day and imposes delay. "Franken-food" characterization of GM foods reinforces consumer wariness. Over time, however, tempers may wane—especially if time and research can provide a "clean bill of health."

SUMMARY

Nearly 100 percent of the world's food supplies are likely to be genetically modified in some manner within another twenty years. Over 70 percent of processed foods in the United States already are genetically modified in some manner.

The benefits of advances are likely to prove too important to be discouraged or denied. Complaints about "messing" with nature will fade as new capabilities for "perfecting" the foods are realized.

Former President Jimmy Carter put the matter aptly: "Biotech is not the enemy. ... Starvation is." Genetic technologies hold promise to provide enough food—at affordable costs—for everyone. What will the outcome be? It's up to all of us—each in our own way.

REFERENCES

Charles, Daniel. *Lords of the Harvest: Biotech, Big Money, and the Future of Food*. Cambridge, Mass.: Perseus Publishing, 2001.

Leisinger, Klaus M., Karin M. Schmitt, and Rajul Pandya-Lorch. *Six Billion and Counting: Population and Food Security in the 21st Century*.

Washington, D.C.: International Food Policy Research Institute; distributed by The Johns Hopkins University Press, 2002.

McHughen, Alan. *Pandora's Picnic Basket: The Potential and Hazards of Genetically Modified Foods*. New York: Oxford University Press, 2000.

Nestle, Marion. *Safe Food: Bacteria, Biotechnology and Bioterrorism*. Berkeley, Calif.: University of California Press, 2003.

Pringle, Peter. *Food, Inc.: Mendel to Monsanto—The Promises and Perils of the Biotech Harvest*. New York: Simon & Schuster, 2003.

■ Essay Seven ■

BIOLOGY IN YOUR FUTURE: PREPARE TO BECOME SOMEBODY NEW

Melvin Konner, Ph.D.
Samuel Candler Dobbs
Professor of Anthropology, Emory University

Despite the troubles in the world, it's a great time to live. One reason is that the pace of scientific discovery is so much faster than it has ever been. Jules Verne, a French writer who lived in the 1800s, predicted that people would go to the Moon. But we had to wait a hundred years for it to happen, and poor Jules wasn't around to see it.

Today you can hear forecasts that sound weird but may come true in ten years. Do you know what a record is? Maybe, but only if your parents saved some from long ago. Your kids won't know what a CD is, and you won't save them, since all the CDs ever made will fit on a single handheld hard disc. It will take a second to download an album, and albums will be so cheap they won't be worth stealing. In fact, in a few years, you will be downloading music videos to your cell phone in seconds flat and sending them to that special person—introduced, of course, by a video of you saying, "How cool is this?"

So here are some forecasts about how your life may change within ten years because of discoveries in biology. Not all of them will be right, but most of them will, and if you want to, you can help bring them about sooner. These advances will change the way all of us live, love, and think about ourselves.

MEDICINE AND DISEASE
We are smack in the middle of two great scientific revolutions that are changing how we fight diseases. One is genetics; the

other is imaging. The human genome was sequenced in 2000, and thousands of scientists in every corner of the world are nailing down what each gene does. Patients with some rare diseases are already being helped. But all of us will benefit in the next ten years as the genes contributing to our most common killers—heart disease, cancer, diabetes, and others—are analyzed.

Some will be fought by changing the genes themselves. For example, we will strengthen the genes that keep some cancers from growing. Others will be cured by inventing new drugs to change a step in the chemical path to the disease. And you will be tested for how likely you are to get these diseases long before they start. Then, you will get health advice tailored to fit your genes. Are you someone who should hardly ever eat fatty food? Or are you more resistant to fat but sickened by too much salt? You will be able to go to your corner lab and find out.

As for imaging, we can already look inside the body and the mind in ways that few dreamed of a decade ago, and progress right now is incredibly fast. Cancers or infections that would have been missed a few years ago light up like colored bulbs on new kinds of scans—all done from the outside, without dangerous rays, while you lie comfortable and awake. Electrical impulses racing around your brain can be followed in detail to find out where your headache or brain virus or even blue mood might be starting. In the next ten years, smaller and smaller things will be seen in safer ways, and the gains in predicting and controlling disease will be huge.

And that's just for the next ten years. There are even more weird and wonderful things around the next corner. How about going into the bathroom and seeing the wall light up with medical information and advice:

"Getting low on calcium. Drink some skim milk or take a pill."

"Bad germs on your hands right now. Wash before you eat or rub your eyes."

"You haven't flossed in days. You want to keep those teeth or what?"

"Lost a pound. Way to go!"

Also on the horizon is the magic of nanotechnology. That means super-small machinery made out of molecules—but not ordinary chemicals; they are much cooler than that. They are assemblies of molecules that work like machines—microscopic ones. Armies of nanoscale robots—nanobots—might scout your bloodstream for bad stuff: the tiniest new clots, for instance, or viruses, or fatty deposits that might harden your arteries. They could then gobble up the gook and break it down into harmless parts that would flush safely out of your system. Finally, they would do the same to themselves—I mean, you don't want microscopic robots hanging around inside you! They will simply do their job and disappear.

Eventually you will be able to add computer chips to your brain. Scientists are making chips and brain cells talk to each other. Implants help deaf people to hear, and sensors on the chest help blind people to see. Soon, paralyzed people will control robotic arms with their thoughts. Someday we may be able to hear ultrasonic high-pitched sounds that now only our dogs can hear or "see" in the dark with sonar. Trouble with arithmetic? No problem. One of your simpler chips will calculate square roots in a microsecond and transmit the answer right into your brain as if you had thought of it yourself. Think of the possibilities for the SAT!

Speaking of the SAT, how about taking a pill that makes you smarter? Many labs are trying to find one. But that's not a treatment for a disease; it's using medicine to improve yourself. If that sounds improbable, it's not. We are already doing a lot of it. Growth hormone was first given only to short kids with a deficiency. But it was later found to make kids taller even if they have normal levels of the hormone, and some kids get it just for being short. New pills help many men who suffer from impotence, but others are using them just to make sex better. We have pills for depression, but where is the line between depression and blue moods?

The line keeps moving, and many people are now taking antidepressants because it makes them more energetic and happier. Finally, the prescription stimulant that you or someone you know is taking is officially given only for attention deficit disorders, but that category has grown to include millions of children.

Is that bad? It's a question we need to think about a lot right now, because in the next ten years, new medicines will be developed in each of these areas and in others we haven't thought about. Bummed about being dumped in a relationship? We may have a pill for that. Can't get your boyfriend to commit? Research on fidelity in animals may lead to a pill for that, too. Your girlfriend isn't as interested in sex as you are? Someone may invent a drug to change her. Steroids are dangerous and illegal in organized sports, but we may have new versions that are safer. If you are involved in a sport where it is not illegal to do so, will you be able to resist taking medicine to make you stronger and faster? Endless possibilities and even more questions.

DIET AND WEIGHT

Every issue of every self-help or beauty magazine has diet advice, but the same things are just repeated in slightly different ways. Fads come and go, but the weight stays on, and eating disorders like obesity, anorexia, and bulimia are more and more common. A few imperfect chemical fixes have been tried, but they are just the beginning. Science is answering key questions: What makes us hungry? Why do we eat when we don't need food? Why do we love the four "basic" food groups—sweet, salty, fatty, and alcoholic?

We love them because the taste for them is built in. Our distant ancestors, who ate wild plants and animals that they had to get themselves, had much trouble finding these things. Sweet and fatty foods were important because they helped keep us from starving, while a certain amount of salt is necessary for life. Alcohol and other addictive drugs like nicotine and cocaine blindsided the human brain when people first discovered them. Nothing like them had ever entered the brain before, and we enjoyed the effect—pleasure in the short run leading to great pain.

Willpower makes a huge difference, and a lot of us can "just say no." But not everyone can—especially to food—and those who cannot may be able to get help. Prescription drugs will turn off hunger in the brain before we eat too much. Other pills may derail digestion after we gorge—but safely, without side effects. Will this make our willpower even weaker? Maybe, but the pills won't work perfectly, and there is no substitute for gaining control over your own life—with a little help from your new molecular friends.

GM (NO, NOT THE CARS)

People in some countries have opposed genetically modified (GM) foods, but they are quickly being used throughout the world. Americans came first: We're a practical people, and if it works, we use it. But where such foods will make the most difference is in the developing world. New genes slipped into wheat and rice are making the land more productive and the grains more nutritious.

We will have fruit and vegetables fortified with even more vitamins than they already have. Also, the produce will be protected against insects and plant diseases. Worried about the health effects? There are no proven ones, although some skeptical scientists have been looking for them for years. GM foods that protect and preserve themselves will be safer than foods sprayed with chemicals, as most are today.

At any rate, after centuries of breeding and improving plants, we modified the genes in them through artificial selection. The grapes we eat today are seedless and much larger than they were a century ago. This involved slowly changing their genes by selecting and grafting more desired varieties. Now we do the genetic modification faster, more efficiently, and in many different ways. The food you put in your body ten years from now may be as different from today's food as it is from what your great-grandparents ate.

Animals, too, will be changed by genetic modification, just as they always have been by artificial breeding. Over the centuries we made beef cattle much fatter, with more "marbled" meat. This meant that the worst kind of fat, saturated fat, was laced through your steak or burger. Sure, it tasted great, but it also helped cause an epidemic of heart disease by hardening our arteries. And it is making Americans more obese every year, which means an increase in diabetes. But new experiments have changed the genes of some animals to make their fat mainly *un*saturated—more like the fat in almonds. That will reduce heart disease—although we will still have an obesity and diabetes epidemic unless we eat less.

Finally, in a few years, we will have "libraries" of GM bacteria to use for a thousand purposes, from cleaning up oil spills to making medicine to creating better conditions inside our intestines. Controlling these invented microbes will be challenging, but it can be done.

STDS AND PREGNANCY

Some experts say that the two most important biological advances of the nineteenth century were the germ theory and the perfection of rubber manufacturing. The germ theory led to enormous benefits through sanitation, vaccination, and treatment of infections. This tipped the scales in our favor in battling our oldest and worst enemies. But why rubber? This made possible the development of a device that reduces the risk of pregnancy and sexually transmitted diseases: the condom.

No doubt the best way to avoid these unwanted outcomes is not to have sex until you are ready to commit to one person. An amazing trend has worked its way through the teenage world in the last decade: a great decrease in casual sex and unwanted pregnancy. But as with the problem of obesity, the flesh is often weak. If you are going to have sex, condoms are the best way to reduce the risk of pregnancy and STDs.

But what about ten years from now? Better condoms will be safer and easier to use. New medicated gels will add another safety factor by killing germs and preventing the sperm from finding the egg. Couples committed to each other will be able to have sex without the risk of an unwanted pregnancy and without the fear that a mistake one of them made before they met could infect the one they love. We are always in a race against bacteria and viruses, and they evolve constantly to counter our best efforts to defeat them. AIDS, believe it or not, was something new in 1980. Sex is a favorite human weak spot in which germs will always take advantage of us. But new medicines and new preventive strategies—abstinence among others—will help us keep them under control.

BABY-MAKING, BABY-CHOOSING

What about when you actually *want* a pregnancy. You may find it is not as easy as you thought, and you will turn to reproductive technology for help. This may be as simple as putting more concentrated sperm into a womb. Or it may involve surgery to take egg cells from a woman's ovary, put them together with sperm in the lab, and place the developing embryo back in the womb. This is called in vitro fertilization, or IVF, and it leads to what are called "test-tube babies." This term made a lot of people afraid when IVF was first used years ago, but the first baby born that way, Louise Brown, is now grown up and perfectly normal. More than a million babies worldwide have followed in her footsteps.

We can check the embryos for genetic diseases before they are put in the womb. This can be done when there are only eight cells. If a genetic disease is found, that embryo is not used. As an ordinary pregnancy progresses, some cells can be taken out and checked. This can be another way to prevent the birth of a child with a terrible disease. Over the next few years, this checkup will be done earlier in pregnancy and thousands of diseases will be tested for.

This sounds good, but once you have this kind of control, many other things become possible. People can already choose the sex of their baby. In ten years your doctor will be able to identify not just severe childhood diseases but *tendencies* toward adult ones, like diabetes or cancer, that don't even appear until middle age. It will become possible to test for things that have nothing to do with disease, and we may be able to *change* the genes in the womb.

Maybe you want a baby with silky black hair or one with thick blond curls. How about one with a high IQ, instead of the average-IQ baby the doctor says you are carrying? Is your baby destined to be short? How about putting some genes for growth factors into him and maybe getting an NBA player? Or something creative, like a poet or a composer? No wait, those creative people have a ton of emotional problems. Let's make sure he's *not* creative.

You get the idea. But that's not all. We hear news about cloning, and many different animals have been cloned from an adult animal's cells. One wealthy woman made a large donation to a lab to get it to try to clone her dying dog. No one has done this yet with humans, but they are certainly trying. Within ten years, they will probably succeed. That means you will be able to have a baby who's basically your identical twin.

Why would anyone want that? Vanity, maybe, or intense curiosity about what would happen if the tape were played over again with different parents. Actually there are a lot of differences between identical twins—or clones—because of chance events as the human embryo assembles itself in the womb. But one way or another, some people are going to try this. It will be up to your generation to decide what to allow and what to prevent, because it *will* be possible.

LIFE ELSEWHERE

The chance that Earth is the only place with life is incredibly small, and pretty soon we will prove that it isn't. The first planet sailing around a star other than the Sun was just discovered a few years ago, but hundreds more have been found and millions more exist. Thousands of amateur astronomers with backyard telescopes have joined a small army of stargazers, looking for faint changes in light that mean a planet is passing by. (Join up, and you could have a planet named after you.)

In ten years, though, we won't be able to get to even the closest star. So for now we'll have to settle for finding life right here in the solar system. Where? Well, we're now sure that Mars has water. It's mostly ice, but it once was liquid, and that means Mars probably once had life. It might still. Titan, a moon of Saturn, is covered with oily liquid—rivers of gas-like substances that might support some different kind of life we don't understand. And most likely of all, Europa, a huge moon of Jupiter, is covered with regular old ice, and beneath that frozen surface is a vast ocean of salt water—exactly the kind of environment in which life on Earth began.

You can judge for yourself how finding life on another planet or moon will change how we think about life on Earth, because it will probably happen in the next ten years.

LIFE HERE

All life on Earth is threatened by one scary species—us. We kill other species by the thousands by burning and cutting down forests, damming up rivers, pouring oil and other junk into the oceans, and heating the Earth by burning oil, coal, and gas. There are already too many of us, and our numbers will go on growing until around 2050.

The good news is that the next ten years will see the beginning of the end of this trend. China and Italy have arrived at a one-child family in different ways, and many other countries are following their lead. Women in Bangladesh, one of the poorest countries, have started to have smaller families much sooner than experts predicted. This is part of a trend that proves an important recent discovery: One of the best things you can do for a poor country is send girls to school and keep them there longer.

Educated girls grow up to have safer childbirths, healthier children (and husbands!), a higher standard of living, and better hopes for the future. Because of these changes, they have smaller families. In the next decade, the impact of this trend will make it clear to all that the human population *can* be controlled and even decrease—without forcing people to do it. Then we can concentrate on creating a better life for those who are here and on protecting the huge variety of life on our planet so that elephants, whales, tree frogs, eagles, and hummingbirds will still be around for your grandchildren to marvel at and enjoy.

UH-OH

Does all this scientific progress have a down side? Of course, advances always do. Anything with the power to help has the power to harm. Diet drugs might weaken our will and make us dependent. Easy protection against pregnancy and STDs might make us promiscuous. Certainly the ability to control what kinds of genes your baby has brings up all sorts of ethical questions that won't be easy to answer. Some GM food might slip by the health watchdogs and do some sort of unplanned harm. Even the search for life on other planets could give us a false hope of leaving Earth and undermine our will to protect what we have here.

But this is nothing new in human life. Stone tools could be used to build or to kill, fire to warm or destroy, words to tell the truth or lie. New technologies like the steam engine, the automobile, and the computer all have been used for good and evil. No one can stop the evil uses completely. But one thing has never happened in all of human history: People have never stopped moving forward in science and technology. This will continue, and like all the people who have come before us, we will have to figure out how to use the advances for good.

FURTHER READING

Greenfield, Susan. *Tomorrow's People*. New York: Penguin/Putnam, 2003. Gives a wonderful take on how biological and medical advances will change our lives. Aldous Huxley's *Brave New World* (any edition) remains a great story and a highly readable warning about how our new level of control over biology could go wrong.

WEB SITES

www.mic.ki.se/Diseases/Mednews.html
An advanced Web site for following what's new in the science of medicine, with links to many other free sites.

82 MOVING ALONG: FAR AHEAD

www.nytimes.com

A great source for news about advances that are changing our lives. Click on "Science" or "Health" to get all the articles from the past week for free.

www.quackwatch.org/index.html

This site will help protect you from the medical scams and false claims that are so common these days.

www.worldwatch.org

Probably the best, most up-to-date, and most readable accounts of the state of the environment.

Part Three

NANOTECHNOLOGY—TOMORROW

We must see to make what happens,
in substantial part, a logical
result of our having planned
that it should happen.
—Simon Ramo,
Century of Mismatch

Seemingly limitless possibilities are opened up by our new-found ability to manipulate atoms one at a time, our ability to transform matter at the subatomic level. Known as nanotechnology, it has one forecaster contending that "even Alvin Toffler [author of the 1970 best-seller, *Future Shock*] could not have envisioned the tidal wave of change that will hit us when nanofactories make the scene."[1] Venture capitalists are falling over one another to get in early; governments are rushing to finance huge research and development programs, lest their national economies fall far behind; and businesspeople wedded to pre-nano realities are scratching their heads and worrying about a technology that looks likely to change everything soon, and not necessarily in their favor.

Breakthroughs, for example, are expected from the convergence of nanotech, biotech, infotech, and cognitive science. For one, enthusiasts expect that a new specialty, *neurotechnology*, may make it possible to manipulate the brain as never before. Data from nanobiochips that analyze DNA, RNA, and proteins may be combined with data from next-generation brain imaging systems, creating new tools for mental health that could make bioanalysis easier and cheaper. This could mean dramatic cures for mental illness and related mental stresses. Problems, however, may arise from uses in law enforcement or the military (neuroweapons).[2]

The first essay explores the potential of all of nanotechnology, with care taken to note perils as well as potential rewards. The second essay comes at the topic from a unique and valuable angle, asking how nanotechnology might radically challenge conventional systems of economics. It remains possible that in your lifetime capitalism will have to accommodate a nano impact of an unprecedented nature. As the second essayist explains, nano may yet have the ability to help us finally end "the age-old triangle of want, fear, and ignorance," arch components of traditional economic systems. And, as a concerned forecaster contends, this future is "very near. Much nearer than we might have thought."[3]—Editor

NOTES

1 Mike Treder, "Molecular Nanotech: Benefits and Risks," *The Futurist*, January–February 2004, 42.
2 Zack Lynch, "Think Nano Has Ethical Problems? Just Wrap Your Brain Around Neuro," *Small Times*, 5 March 2004. www.smalltimes.com/document_display.cfm?document_id=7522
3 Treder, *op. cit.*, 46. "Will it be easy to realize the benefits of nanofactory technology while averting the dangers? Of course it will not. Is it even possible? It had better be." *Ibid.*

■ Essay Eight ■

NANOTECHNOLOGY: BIG REVOLUTION WITH SMALL THINGS

Jim Pinto
Technology Futurist

What if you could take the tiniest specks of matter, atoms, and molecules, and make them into intelligent machines? That mind-boggling idea has fascinated scientists for decades. Now, efforts to make this real are accelerating. An awesome revolution is rushing your way—one called nanotechnology—and it is arguably more incredible and far-reaching than anything ever before. As its impacts are very likely to be felt as soon as the next ten to twenty years, you will undoubtedly feel its effects, for better and for worse.

In less than ten years, nanotech is very likely to have a huge effect on many industries, including manufacturing, health care, energy, agriculture, communications, transportation, and electronics. Within a decade or so, nanotech could be the basis of $1 trillion worth of products in the United States alone and could create 2 million new jobs. The sooner young people think creatively about how nano may revolutionize their lives, the better.

BACKGROUND

More than forty years ago (December 1959), at the California Institute of Technology, Richard Feynman gave a talk entitled "There's Plenty of Room at the Bottom"—on the challenge of manipulating and controlling things on the atomic scale. His talk inspired scientists to begin creating molecular devices that could compute, assemble, and replicate themselves. The main idea was that atoms could be treated discretely to build structures, and matter could be manipulated into tiny machines capable of self-replication.

In 1986, K. Eric Drexler published *Engines of Creation*, a groundbreaking book. He coined the word—nanotechnology—which uses the Greek word *nano* (dwarf). It refers to a measurement called a nanometer, or one-billionth of a meter, the width of about four individual atoms (or one 80,000th the width of a human hair). Drexler described ways to stack atoms, assemble machines much smaller than living cells, and make materials stronger and lighter than anything dreamed of today. We began to think we could assemble myriad tiny parts into intelligent machines.

The Holy Grail of nanotechnology is self-assembly, which may soon be an effective nano-engineering tool. Self-assembly is nothing new: biology does it all the time; in chemistry, molecules team up to form structures. Indeed, the concept of self-assembly grew out of attempts to aggregate molecules spontaneously into specific configurations. Now, nanotech self-assembly is attempting the same.

When we get down to the nanometer size, the classical laws of physics change. Once atoms can be manipulated, it is possible to produce new materials with desired properties: smaller, stronger, tougher, lighter, and more resilient than anything that has ever been made. It becomes possible to build useful objects atom by atom, like electronic circuits thousands of times more powerful than today's silicon chips. And there are also exciting medical implications. If atoms can be manipulated one by one, then it might be possible to edit DNA—for example, to prevent certain human diseases and worse aspects of the aging process.

AN INFLECTION POINT

Many serious forecasters—including this one—believe that nanotechnology is *the* next big revolution, an "inflection point" that will change almost everything! It is likely to prove bigger than the Internet and more far-reaching. It will destroy a lot of business dinosaurs and create vast new wealth. It is developing to the extent that practical applications are already here, and a surge of new materials, products, and systems will be coming within the next few years, and certainly within the next decade.

As nanotechnology advances into practicality, achievements will transcend and unite such diverse sciences as physics, chemistry, biology, and even computer science. Soon, vast sectors of the economy will be affected, from biotechnology and health care to energy and manufacturing. Nanoscale engineering is already translating science to practical design and assembly processes. Nano, in short, is likely to shake up just about everything on the planet.

NANOTECH APPLICATIONS

In 2003, *Business Week* named nanotechnology as one of the "Ten Technologies That Will Change Our Lives." The commercial interest in nanotechnology is being driven by visions of a stream of new nanotech products and applications that will lead to a new industrial revolution in which almost every industry is likely to be affected.

Unlike the Internet, nanotech has the potential for creating entirely new materials, products, systems, and markets. Atoms and molecules can be manipulated to create useful materials, devices, and systems: materials 100 times stronger than steel but lighter than plastic, super drugs that eradicate cancer cells without side effects, and self-assembling minibots that can reproduce any substance at the atomic level.

On the most mundane level, an example of nanotechnology's gains involves scratchproof glass and tiles that shed dirt and never need cleaning. More sophisticated applications include precision drug-delivery systems that can be swallowed like pills or computers the size of a sugar cube that could hold the entire contents of the Library of Congress.

The biggest impact is likely to come from nanoelectronics. The promise here is smaller, faster, and cheaper products than conventional approaches could ever achieve. Advances have come with remarkable speed. As recently as 1998, researchers struggled to rig up a single nanoelectronic component: a molecule that acted as a rudimentary switch. Today, they are connecting dozens of these nanoscale components and are looking to assemble entire devices, like memory chips.[1]

Nanotech will undoubtedly be used to create several low-cost, highly accurate biological and chemical sensors. NEMS (nano-electro-mechanical systems) are quickly becoming practical, bringing ultra-sensitive sensors and ultra-strong actuators that might replace damaged human tissue or power tiny robots.

Tiny sensors and actuators are already everywhere—triggering airbags, controlling colors in inkjet printers, and projecting light for digital cinema. These are MEMS (micro-electro-mechanical systems) that use the same fabrication methods as silicon chips. Now the next step in miniaturization is NEMS. There are huge applications for tiny, inexpensive nanosensors—from medical diagnostics to chemical and biohazard detection to vast arrays of wireless networks.

Companies are already commercializing these products. Applications have included better "skins" for aircraft and automobiles, and tiny devices that can travel along capillaries to enter and repair living cells and help make the human body stronger. Nanotechnologists are also developing a polymer "glue" to persuade hundreds of thousands of tiny nanoparticles to group together into large, highly ordered structures. This allows single atoms to be pushed around and positioned—used, for example, to edit DNA. Each of the tiny particles could be a memory element in a molecular computer.

Commonly used materials take on entirely different characteristics when "assembled" at a molecular level. So, even today, the biggest nanotech market is in materials. The use of nano-size particles in products like cosmetics, sunscreen, paints, and a host of other products is already commonplace. Many specialized nanotech start-ups are emerging and whole new industries will grow up around them.

Right now, for example, there are many start-ups selling carbon nanotubes, the strongest and most conductive fibers known.[2] Molecule-size components are being assembled into complex composites and "smart" materials. For example, nanostructured membranes are being developed for efficient filtering of pollutants from water or air. Buildings and machines could signal when they need maintenance—and perhaps repair themselves. Our clothing could monitor our health and alert us to environmental hazards.

Against the backdrop of the war on terrorism, work is progressing on a nationwide sensor network that someday could provide a real-time early-warning system for a wide array of chemical, biological, and nuclear threats across the United States.[3] With a $1 billion budget in 2004, the U.S. Department of Homeland Security is doing a significant amount of new development, plus coordinating the efforts of key scientists at national labs. The core technology relates to materials, sensors, networks, and chips. Field trials of prototype networks are starting.

Terrorism, however, is not the only incentive. The ultimate vision for sensor nets is to make them smart, autonomous, and self-aware. Imagine logging onto the network and typing in, "Does my lawn need more water?" The sensor network would examine figures from moisture sensors around your home and send back a prompt reply. At warehouses, managers could quiz shelf-mounted sensors about inventory.

NEAR-TERM NANOTECH

Some of these nanotech materials and products are still in the laboratory, just a future vision.[4] How about the near-term? Here are just some of the real nanotech products that are on the market:[5]

* Sunscreen makers have found that nanoscale particles cover the skin more thoroughly and do not reflect light. So, Procter & Gamble is adding nanosize particles to its sunscreen lotions, with significantly better results.
* With a rubber core that uses tiny "nanoclay" particles to form an airtight seal, Wilson tennis balls retain their air pressure twice as long as ordinary balls.

MOVING ALONG: FAR AHEAD

* Eddie Bauer, Lee Jeans, and others are selling stain-free and wrinkle-resistant slacks developed by Nano-Tex. Billions of tiny whiskers create a thin cushion of air above the cotton fabric, smoothing out wrinkles and allowing liquids to bead up and roll off without wetting the fabric.

In the next two to five years, we can also expect:

* Car tires that require air just once a year.
* Self-assembly of small electronic parts.
* Artificial semiconductors based on proteins.
* Instant, error-proof pregnancy tests.
* Medical diagnostics computer chips.
* Portable concentrators that produce drinking water from air.

On the horizon, in five to ten years, will be:

* Erasable, rewritable paper for books and newspapers.
* Bulletproof armor.
* Ultra-light, ceramic car engines.
* Voice-recognition hearing aids.
* AIDS and cancer treatments.
* Smart buildings that resist earthquakes.

With nanotechnology, today's supercomputer could become tomorrow's wristwatch PDA. All of these marvels, and many more, are scientifically possible. The difficulty is to figure out how and when these things will happen.

DEBATING NANOTECH
In 1986, with the publication of his groundbreaking book, Eric Drexler warned: "There are many people, including myself, who are quite queasy about the consequences of this technology for the future. We are talking about changing so many things that the risk of society handling it poorly through lack of preparation is very large."

In his famous article, "Why the Future Doesn't Need Us," Bill Joy, the chief scientist and cofounder of Sun Microsystems, and the coauthor of the popular Java programming language, wrote eloquently: "Our most powerful twenty-first-century technologies—robotics, genetic engineering, and nanotech—are threatening to make humans an endangered species."

Many prominent "ethicists" join Bill Joy in calling for a technology moratorium. The problem is the inexorable quest for progress, with impatience that causes incomplete knowledge and understanding of the complexities of life. The message is that technology, irresponsibly applied, spells danger and even threatens death and/or extinction.

Freeman Dyson, an emeritus professor of physics at Princeton, takes the opposite, libertarian view. In "The Future Needs Us"—his response to Bill Joy's article—he argues that the dangers do not come from any particular technology (like nanotechnology), but from the possibilities of moving forward without fully understanding the consequences. Risks are unavoidable, and no possible course of action can eliminate all risks. Action must be based on balancing risks and costs against benefits, including the protection of human freedom to explore.[6]

CONCLUSION

Nanotech is a revolution that is fast approaching. The U.S. National Nanotechnology Initiative puts nanotechnology at the top of the U.S. science and technology agenda, with an estimated $961 million in funding in 2004. In addition, it is estimated that countries worldwide are investing $2 billion in this field.[7]

Within a decade or so, nanotech will have huge effects on manufacturing, health care, energy, agriculture, communications, transportation, and electronics; it will be a $1 trillion business and could create 2 million entirely new and exciting job opportunities.[8]

You are indeed fortunate to be at the age when you are

coming out of school and perhaps entering college, just as a new technological revolution is about to unfold. It's important that you think about how nanotechnology can affect your life, how you can help contain the perils it may pose, and benefit from the new and wonderful things it may yet create.

ADDENDA: FICTION PORTRAYS A DYSTOPIAN FUTURE

In his 2002 techno-thriller *Prey*, Michael Crichton, the author of *The Andromeda Strain, Jurassic Park,* and other best-sellers, weaves a story about the perils of nanotechnology. This is combined with a technically realistic account of distributed intelligence, self-organizing systems, and emergent behavior.

Crichton doesn't allow the reader to relax with the feeling that the danger is fictional—the book includes an introduction to emphasize the direct link to reality.

Drexler's warning (above) is included, as well as an extract from a Santa Fe Institute paper by J. Doyne Farmer and Alletta d'A. Belin, "Artificial Life: The Coming Evolution." Here is that excerpt:

> Within fifty to a hundred years a new class of organisms is likely to emerge. These organisms will be artificial in the sense that humans will originally design them. However, they will reproduce, and will evolve into something other than their initial form; they will be "alive" under any reasonable definition of the word. The pace of evolutionary change will be extremely rapid. The advent of artificial life will be the most significant historical event since the emergence of human beings. The impact on humanity and the biosphere could be enormous, larger than the Industrial Revolution, nuclear weapons, or environmental pollution. We must take steps now to shape the emergence of artificial organisms; they have potential to be either the ugliest terrestrial disaster, or the most beautiful creation of humanity.

NOTES

1 Some features of computer chips are already pushing past the 100-nanometer mark, into the nanotech realm. Pentium 4 processors with 90 nm transistors are in production. A nanotech project at Intel is developing a tri-gate 3–D-like cell that could lead to terahertz transistors (1 trillion cycles per second). The race is on to build computer circuits from molecules, so that more and more components can be crammed into less and less space, using less and less power.

2 Carbon nanotubes are tiny tubes about ten thousand times thinner than a human hair. They consist of rolled-up sheets of carbon hexagons. Discovered in 1991, they have the potential for use as minuscule wires or in ultra-small electronic devices. They can now be grown on silicon wafers, creating the possibility of combining nanoscale circuits with conventional chip manufacturing. The huge markets involved have inspired many teams of researchers to explore nanometer-scale structures, looking for new ways to manufacture circuits from single molecules.

3 IBM has shipped more than 5 million disk drives with a new nanostructured magnetic coating (referred to as "pixie dust") that quadruples the data storage. But, as the components shrink, manufacturing costs increase, and there are physical limits to the minimum size of a useful silicon transistor or the data storage density of a magnetic disk. So, totally new nanotech devices are being explored to process and store information.

4 Carbon nanotube transistors, for example, can be made smaller than any possible silicon transistor, with far better performance. And other new ways of storing information are also being explored. A nanomechanical system called "Millipede" stores data as tiny, erasable indentations in a thin plastic film; this could allow trillions of bits of information to fit within a chip that could be used in a wristwatch PDA.

5 William Atkinson makes his living by explaining technology to business types. He demystifies nanotechnology and describes the science and business behind it in his new book: *Nanocosm: Nanotechnology and the Big Changes Coming from the Inconceivably Small* (American Management Association, 2003). In layman's terms, he discusses the complex science and its real near-term

applications in manufacturing, pharmaceuticals, information technology, and many other markets. I draw here on Atkinson's list of products and applications that are coming short-term. Another new book I recommend: *The Next Big Thing Is Really Small* (Crown Business, 2003) by Jack Uldrich with Deb Newberry provides a good, introductory explanation of how atoms and molecules are manipulated to create useful materials, devices, and systems.

6 And clearly, uncontrolled activity will progress (in uncontrolled nations) no matter what.

7 Numbers quoted on the National Nanotechnology Initiative (NNI). Web site at www.nano.gov/.

8 *Ibid.*

FURTHER READING

Drexler, Eric K. *Engines of Creation: The Coming Era of Nanotechnology*. New York: Bantam, Doubleday, Dell, 1986. This book started the revolution, and Drexler is now considered the father of nanotechnology. The book is easy to read—read it!

Dyson, Freeman. "The Future Needs Us": www.nybooks.com/articles/16053.

Engines of Creation: www.foresight.org/EOC/ Take a look at the Web version of Drexler's book. Here is a link to reviews and where you can buy a copy: www.jimpinto.com/reading.html#DREXLER.

Joy, Bill. "Why the Future Doesn't Need Us." *Wired*. April 2000. Bill Joy's famous article can be found at www.wired.com/wired/archive/8.04/joy.html.

"The Once and Future Nanomachine": http://www.ruf.rice.edu/~rau/phys600/whitesides.htm.

"Self-Assembling Devices at the Nanoscale": http://www.nsf.gov/od/lpa/news/03/pr0377.htm.

■ Essay Nine ■

... AND THE BUBBLEGUM POPS: NANOTECH VS. CAPITALISM

Glenn Hough

If at first the idea is not absurd, then there is no hope for it.
—Albert Einstein,
as quoted in Des MacHale's *Wisdom*

Any useful statement about the future should seem ridiculous.
—James Dator,
Dator's Law for Future Studies

If I tell you something ridiculous about the future, will you believe me? Perhaps in light of the above quotes you should.

The Dator quote is one of the more useful axioms in future studies. Its validity can be verified by a simple thought experiment. Just ask yourself: What would a person, who was your age in 1804 or 1704, think if you could tell him/her about your life? Just your ordinary average daily life that you take for granted and never have to think about. What would they think?

Chances are they would think you were a skilled liar, an escaped lunatic, drunk, or the recipient of a vision. But they wouldn't think you were telling them the truth. That's how much our society has changed in a few hundred years. Our society, from their point of view, is ridiculous. Women voting or holding professional jobs outside the home? Preposterous.

Interracial marriage and no slavery? That's crazy talk. Men on the Moon? What have you been drinking?

So, if I tell you something preposterous, ridiculous, crazy, or absurd about nanotechnology's future, can you believe me?

We have heard all the hype about nanotech. It'll be bigger than the Internet. It'll change everything; think of the jobs, they say. Nanotech is a hot topic, be it K. Eric Drexler, who produced the seminal work, *Engines of Creation*, or the plethora of articles written for *Wired* magazine, or the growing collection of features in the *New York Times* and the *Washington Post* on the billions of dollars that investors are pouring into nanotech research and start-ups. But where they stop is like building a house; stopping at the foundation and saying, well, that's it, done. When the concrete is poured for the foundation, you are approaching the middle point of the project, with a lot of heavy work left to go.

So, are you ready for something ridiculous? Are you ready to glimpse the potential of the completed house of nanotech?

Since the dawn of humanity, we have been plagued by three great evils: a reinforcing triangle of want, fear, and ignorance. From this triangle, all the other human miseries flow: war, pestilence, and so on. Nanotechnology is the key to eliminating want from the human condition. And if want falls, if one corner of the triangle falls, the other two will be taken down with it.

It's preposterous, isn't it? But that's the potential; that's the part they don't tell you about. It's really quite logical. What is the difference between sand and water? Stone and wood? It's molecular structure, or more precisely, it's sub-atomic structure—the pattern the atoms form, which forms the substance.

What is nanotechnology? The ability to make an object at the atomic level, atom by atom. The ability to manipulate the pattern directly, to take raw material, change its pattern and turn it into something. That something being whatever we wish.

What if we had a vat containing two tons of raw material, amino acids say, and we tossed in a piece of the best Texas-bred steak you could get, about a pound of it? What if nanotech could then use that one pound of beef as the seed, the pattern for the whole? And out of that vat would come nearly two tons of prime Texas beef indistinguishable from the original seed, the original pound, since the technology would manipulate the amino acids—the raw material—at the atomic level to form exactly what we wished. Such technology would eliminate most of the need for cattle ranching, with just a handful of ranches able to supply everybody.

What if we had a machine that could sit on a sand dune in a desert, use solar energy for its power source and the sand it sits on for raw material, and do nothing but produce pure water? If you had enough industrial-size models, making a lake in the middle of the Sahara wouldn't seem such a hard thing to do.

And if making a lake in the Sahara is possible, then meeting the world's water demands seems like something one would have accomplished earlier. Under such conditions, the multibillion-dollar-a-year business of bottled water could dry up. If we had the nanotechnology to make water from sand, then simply filtering existing water would be infinitely easier.

What could happen to the whole capitalist system under such conditions? Want, that is, scarcity, is one of the fundamental pillars of the whole system. It's the prime variable that all economists are taught; all of their equations are based on. The system is built on that assumption: Remove scarcity and the capitalist system falls.

This is evident in today's practice of paying farmers not to grow crops, or the practice of dumping crops or products, so that scarcity, which is called market price, can be maintained. There must always be scarcity in the capitalist system, so whether it's John Steinbeck writing about 1930s sharecroppers, or those of us in the poorest places in America today, scarcity is always maintained. There is more than enough food grown, or that could be grown, so scarcity is now on the money side of the equation these days.

And if nearly everything in the marketplace can be produced by nanotechnology in quantities large enough to supply everyone, will people be paid not to use products to maintain the marketplace, or could the whole system fall and be replaced by something else?

How much do medieval economic structures influence today's economics? There is a trace left, but technology has made the underlying assumptions, the relationship patterns of medieval economics, mostly obsolete.

There is no reason to think that cannot happen again with the current economic system, though the forces that currently benefit will fiercely fight to preserve their privileges, just as the old lords fought the changes springing from the Renaissance. In other words, fear and ignorance, as represented by our current assumptions and patterns of behavior, will fight to maintain the triangle of want, fear, and ignorance, to maintain the status quo.

Nanotechnology can change the underlying assumptions our system is based on; it can change the relationship patterns. Economics as we know it could mostly cease to exist. A trace might be left, but we could move on.

And if we were to change our economics, our political and social structures, our domestic patterns, what we consider to be valid in education, and the assumptions about how we relate to each other, all could change. They could fall, like dominoes, one after another. And in the end, the age-old triangle of want, fear, and ignorance would be broken.

It's ridiculous to think of a world without the sort of capitalism we know, isn't it? A world where money does not dominate everything. But humanity at times gives a serious and maybe even an appreciative second thought to what seems at first glance ridiculous—just ask our friend from two hundred or three hundred years ago about what he/she thinks of today's world. It's quite ridiculous after all.

FURTHER READING

Brown, Lester R. *Eco-Economy: Building an Economy for the Earth*. New York: W.W. Norton & Company, 2001.

Cavanagh, John, et al. *Alternatives to Economic Globalization: A Better World Is Possible*. San Francisco, Calif.: Berrett-Koehler Publishers, 2004.

Drexler, K. Eric. *Engines of Creation: The Coming Era of Nanotechnology*. New York: Anchor, 1987.

Hawken, Paul, Amory Lovins, and L. Hunter Lovins. *Natural Capitalism: Creating the Next Industrial Revolution*. New York: Back Bay Books, 2000.

Mander, Jerry, and Edward Goldsmith, eds. *The Case against the Global Economy: And for a Turn toward the Local*. San Francisco, Calif.: Sierra Club Books, 1997.

McDonough, William, and Michael Braungart. *Cradle to Cradle: Remaking the Way We Make Things*. New York: North Point Press, 2002.

Mulhall, Douglas. *Our Molecular Future: How Nanotechnology, Robotics, Genetics and Artificial Intelligence Will Transform Our World*. Amherst, N.Y.: Prometheus Books, 2002.

Nattrass, Brian, and Mary Altomare. *The Natural Step for Business: Wealth, Ecology and the Evolutionary Corporation (Conscientious Commerce)*. Gabriola Island, B.C.: New Society Publishers, 1998.

Rees, Williams E., Phil Testemale, and Mathis Wackernagel. *Our Ecological Footprint: Reducing Human Impact on the Earth*. Gabriola Island, B.C.: New Society Publishers, 1995.

WEB SITES

Creating Livable Alternatives to Wage Slavery
www.whywork.org/
Real Economic Freedom for Everyone
www.abolishmoney.com/
The Redesign of a Culture
www.thevenusproject.com/
Chaordic Commons—New Ways to Organize
www.chaordic.org/

Part Four

SPACE—TOMORROW

*We must never forget
that the human heart
is at the center
of the technological maze.*
—Stephen Barnes,
The Transparent Maze

We have lived in the space age for nearly a half-century now and fully understand that we have barely begun. Around the world, several advanced industrial nations vie with one another with steadily improving launch vehicles, their own commercial satellites, their own spy satellites, and their remote-sensing equipment. The unique and lesson-rich exercise in nation-cooperation, the International Space Station, moves toward completion, even as exciting discoveries from our Mars rover increase longstanding interest in exploring more and more of that remarkable planet (birthplace of life on Earth?).

Four essays in this section make clear how very far we have to go (no pun intended), especially since so much remains open to debate. The first essay makes a strong case for teenagers caring about the topic. The second has the same writer rebut critics among other contributors to this book who took issue with points made in the first essay. The third explains the controversy over what is best to send to Mars—more and improved rovers or eager, highly skilled human beings? The last essay gives us an opportunity to vicariously spend some quality time in a well-settled colony on Mars, the better to ask if we would care to live there. Plainly, we have 101 exciting and often difficult choices to make—and the enormity of space in which to make them and uncover 101 new ones.—Editor

■ **Essay Ten** ■

SPACE: TEENAGERS AND THE FAR OUT

Jeff Krukin

www.JEFFKRUKIN.com

Should we think of space as only a "program"? What sort of space-related careers might you consider? How does space activity connect to our prosperity? What difference might President George W. Bush's 2004 space initiative make? And where does private industry come in, and NASA get reinvented?

WHY SPACE IS MORE THAN A "PROGRAM"

Six of seven Apollo missions that sought to place human beings on the Moon were successful. A symbol of American power, of the strength of capitalism and democracy, and Apollo scored big! By proving we could choose difficult goals and succeed, it addressed our national weaknesses and helped alleviate our self-doubt.

Apollo, however, was an instrument of foreign policy and a weapon in the Cold War. It had us view space activity through the exclusive lens of government programs. We are only now realizing this is not a crystal-clear lens, but more like an old but comfortable pair of glasses—one that has become blurry with age and neglect. And like old glasses, it does not allow us to see all the possibilities.

NASA's Apollo program did *not* extend our economy into orbit. Nor did it persuade us that we might one day go into space—*really* take up residence in Outer Space. Consequentially, insufficient attention has been focused on the full potential of space ever since Apollo.

When we stop looking at space as a "program" and start

thinking of space as just another "place," we discover exciting new personal ways to relate to it. The first step is realizing that space is the vast neighborhood in which we actually live. It isn't a distant and impenetrable domain. It is a mere 62 miles (100 km) above Earth[1] and thus a matter-of-fact continuation of our environment. You don't have to go very far to experience the darkness, the weightlessness, and the magnificence of space as you look down at the rotating Earth and watch your hometown go by.

As you read this, entrepreneurs are busy building reusable space vehicles for commercial passengers (versus government astronauts), foreseeing the day only five to ten years from now when you will be able to fly into space for a cost of only $100,000. Ten years later you may be able to stay at an affordable orbiting hotel. Imagine future high school students spending a semester at an orbiting campus!

CAREER POSSIBILITIES

What are the job possibilities? Numerous and wondrously varied! You could, for example, become the engineer who builds a reusable rocket engine that is more powerful, yet less expensive to operate and maintain than today's space shuttle engines. Or develop space law that accommodates a new space travel industry. Or be the chef who creates gourmet meals for weightless dining. Or design the marketing campaign for a commercial space company. Or be the researcher who invents … what would you like to invent for use in space?

You could be the economist who influences national monetary policies so they support the development of new space businesses. Or be the journalist who writes a Pulitzer Prize-winning story about a family living, working, and playing in space, which touches the hearts of millions. Or be the diplomat who leads the United Nations commission that defines a space resources development plan for the benefit of all mankind.

You could be the chief geologist at the future Asteroid Mining Inc., the first company to capture and own a near-Earth asteroid rich with iron ore. Or organize a conference to define global safety standards for beaming solar energy from satellites to collector farms. Or be the president of a lunar preservation organization dedicated to maintaining a balance between protecting the Moon's environment and supporting mining operations. Or lead missions to third-world nations, teaching people how to blend new space technologies into their cultures.

HOW DOES HUMAN ACTIVITY IN SPACE CONNECT TO HUMAN PROSPERITY?

As an extension of the human economy, space is a key to humanity's well-being. As long ago as 1970, Kraft Ehricke, a lunar colonization visionary, said it this way:

> While civilization is more than a high material living standard, it is nevertheless based on material abundance. It does not thrive on abject poverty or in an atmosphere of resignation and hopelessness. ... Therefore, the end objectives of solar system exploration are social objectives in the sense that they relate to, or are dictated by, present and future human needs.[2]

In a time of global war against Islamic fundamentalists, in the midst of uncertainty created by global economic integration, it is too easy to believe our opportunities are limited and life is only a me-against-you scenario. Space rebuts this thought, as it contains the abundant resources needed for a global economy able to provide widespread prosperity and unlimited possibility—while minimizing environmental damage.

How can we sustain ourselves—the many billions of us—as the population grows and resource requirements increase? Must we limit ourselves to Earth-bound resources and energy? Helium-3 is fuel for fusion reactors. Available in minute quantities on Earth, it is abundant on the Moon. Huge solar arrays in orbit could capture sunlight and beam the energy into Earth's power grids with far greater efficiency than ground-based solar farms. Asteroids containing iron, nickel, and other ores can be maneuvered into orbit and mined. The possibilities are endless!

WHAT IS THE RELEVANCE OF PRESIDENT BUSH'S SPACE INITIATIVE?

On January 14, 2004, President Bush announced a new vision for American activity in space, with challenging goals for exploration of the Moon and Mars. Referred to as "A Renewed Spirit of Discovery,"[3] was this merely a pre-election ploy or will it help create a sustainable Earth-Moon transportation infrastructure?

According to a poll conducted for the Associated Press and reported the day before Bush's speech, "Asked whether they favored the United States expanding the space program the way Bush proposes, people were evenly split, with 48 percent favoring the idea and the same number opposing it."[4] A *New York Times*/CBS News poll on January 18 reported "...58 percent saying that building a permanent space station on the Moon was not worth the risks and costs."[5]

What do you think of the president's space goals? If you view space as just another government program that must compete for limited funds, how likely are you to support new space initiatives? If you view space as vital to our economic and social well-being, how does this affect your support for space activity? If you understand that involving the full strength of the private sector rather than just using the same few aerospace companies can dramatically reduce the cost of space activity, is your attitude changed?

I urge you to consider that most people who have a negative opinion about space activity simply don't want their tax dollars spent on it when there are other pressing needs. They don't realize that space is more than a "program," and that an increase in NASA's budget can be minimal. They don't see how the private sector can create new industries and new jobs as a result of the president's space initiative.

FINALLY, WHAT OF PRIVATE INDUSTRY AND NASA?

We have the most powerful economy in the world, and yet we

are not using this commercial powerhouse to extend the economy into orbit and beyond. NASA has failed to lower the cost of putting payloads in space because our government is not designed to lower the cost of providing products and services. Rather, it is the private sector's profit orientation that may yet take us into space, just as it has led much of the development of our world.

You don't have to look far to discover that President Bush understands this. In the White House document "A Renewed Spirit of Discovery," the very last sentence of the section, "Bringing the Vision to Reality," is quite clear: "Pursue commercial opportunities for providing transportation and other services supporting the International Space Station and exploration missions beyond low Earth orbit."[6] Inexpensive and reliable transportation is the key, and this can only be accomplished when the private sector is involved.

The first step to accomplishing the president's space vision is to create an Earth-orbit and orbit-Moon transportation infrastructure, and operate it in a commercial (for-profit) manner. NASA cannot do this alone. It has tried and failed with the space shuttle.

Did you know that the space shuttle fleet was intended to fly fifty times each year? It never happened; didn't even come close. Did you know that the space shuttle was intended to lower the cost of putting payloads into orbit? Again, it didn't even come close. Government agencies have no incentive to lower the cost of a product or service, nor are they designed to do so. Corporations must do so, or they fail to survive.[7]

NASA does several things very well, like unmanned exploration of the planets, manned missions to distant bodies beyond the interest or capability of corporations or universities, and basic research and development. None of these require NASA to own and operate space transportation, a basic service best provided by corporations. These corporations need a market, and Bush's initiative sets the way for NASA to provide a large part of that market. If you want to fly into space in the next ten years, you are also part of that market.

When NASA purchases launch services from corporations, it doesn't need to spend billions of dollars developing, owning, and operating launch vehicles. Those budget dollars become available for NASA to do what only it can do, rather than be wasted on what others can provide at far less cost.

Properly directed, NASA can help propel the U.S. economy into orbit while it stretches to accomplish the difficult goals given it by President Bush. Corporations can dramatically lower the cost of launches and space vehicle development. Everything we do in space begins with a launch, so lowering launch costs is vital.

Properly directed, NASA can cease doing the mundane (shuttle launches) and return to the exciting (exploring the far frontier). And this isn't an Apollo-style mission, where we abruptly stop. This is about the permanent settlement and open-ended development of space. This is you and your children able to live in a world, a universe, with increasing, rather than limited opportunities.

SUMMARY

When NASA's planetary rover *Spirit* successfully landed on Mars in January 2004, did you feel something deeper than "Wow, that's cool?" For thousands of years humans have pondered their place in the universe. We now appreciate that humanity is no longer constrained by the box we call Earth. We can truly think "outside the box," for the entire universe is finally our home.[8]

What are the possibilities for you? You can explore your own Human-Space Connection™. You can dare to study *both* religion *and* quantum physics in college, and thereby contribute to humanity's awareness of its relationship with the universe.[9] You can create a new comic strip, one that teaches living *without* Earth-restricted boundaries. So much more remains *for you— endless* possibilities, like space itself.

DISCUSSION QUESTIONS

1) What educational experiences would an orbiting college campus provide that are unique to that environment?

2) Will Apollo-style (politically and/or ideologically driven) Moon and Mars missions succeed as they did in the 1960s? Why, or why not?

3) What can you do to become one of the first business (versus government) people working in space?

4) If humans don't bring the resources of space to Earth, how will we sustain a growing population that requires greater amounts of energy, food, water, and all the other goods that we create?

5) Should humanity explore and settle space, or should we remain forever Earth-bound? Why?

6) What is more likely—that you can become a government-payroll astronaut, a ticket-buying space tourist, or a corporate-payroll space employee? Why?

NOTES

1 John F. Graham: http://www.space.edu/projects/book/chapter3.html.

2 The OURS Foundation: www.ours.ch/spacenews.htm.

3 The White House: www.whitehouse.gov/space/renewed_spirit.html.

4 "Public Divided on Bush Space Plan," *Newsday*, 14 January 2004.

5 *New York Times*: www.nytimes.com/2004/01/18/politics/campaigns/18POLL.html.

6 The White House: www.whitehouse.gov/space/renewed_spirit.html.

7 Commercial aviation provides a wonderful example of a successful government-private sector transportation partnership. The government provides the financial, legal, and safety framework and owns the airports. Corporations own and operate the airplanes.

8 Space is a spiritual realm without boundaries, containing and connecting all life. In the television science fiction series *Babylon 5*, Ambassador Delenn stated, "We are the universe, trying to figure itself out." James Redfield's *The Celestine Prophecy* is a tale of global spirituality within the ultimate setting: the universe. A

SPACE: TEENAGERS AND THE FAR OUT   109

bumper sticker I saw years ago said, "God is too big for just one religion." Michael Talbot wrote in *The Holographic Universe*: "Put another way, there is evidence to suggest that our world and everything in it—from snowflakes to maple trees to falling stars and spinning electrons—are also only ghostly images, projections from a level of reality so beyond our own it is literally beyond both space and time." Perhaps you draw inspiration from Christian rock music, Judaism's Kabbalah, L. Ron Hubbard's Scientology, or Buddhism. No matter your beliefs, you are a creature of the universe. The subatomic structure of your body is the same as that of the stars. The elements within you may be found throughout the galaxy. You exist because the universe was here first. There is no separating you from the universe.

9 Michael Talbot, *The Holographic Universe* (New York: HarperCollins, 1991), 1.

ADDENDA TO "SPACE: TEENAGERS AND THE FAR OUT"

Here is feedback provided by a reader of the preceding essay, to which I've replied below.

1) *I don't think the author realizes that NASA achieved its goal (go to the moon) and has never been given another one. The LEO missions have just been filler and not a mandate. There's a real difference.*

I disagree that NASA wasn't given other goals after the lunar landings. Development of the space shuttle and space station were the presidentially directed manned missions for NASA. I do agree that these goals were merely "filler," and unworthy of the capability demonstrated by NASA and the aerospace companies. If implemented properly, President Bush's space initiative gives NASA exactly what it needs, a worthy and challenging goal and a directive to phase-out its "filler" activities that can be handled by the private sector. By these I mean space station resupply missions and some other orbital activity.

2) *I don't think the author knows that for every 1 billion spent on NASA we get 250 billion back in new goods, services, and whatnot.*

While it is difficult to accurately measure the economic return from NASA expenditures, I believe this value is much too high. Using this figure, NASA's FY2002 budget of almost $14 billion would have returned almost $3.5 trillion to the economy, which doesn't seem likely out of an approximately $6 trillion economy. I believe a generally accepted value is that every $1 spent by NASA has returned $6 to the economy. For interesting reading on this, please see: http://cmex-www.arc.nasa.gov/CMEX/data/vse/session2.html

3) *For the employability aspect: I don't think the author knows that many people "cut their teeth" or "earn the strips" doing space exploration and then turn that knowledge back on Earth to help us better understand things here. That's a rather important point.*

Agreed, this is important. I didn't intend to say otherwise in my essay but to emphasize that this value of space exploration can be multiplied tremendously when it isn't limited to the current NASA-controlled paradigm. I believe this comment demonstrates an outdated view of space, that it be seen as separate from what we do on Earth. It shouldn't be a "cut your teeth" there and then do something valuable here scenario. Instead, what we do on Earth and in space is all part of the same thing; humans living, working, and playing in their environment.

4) *President Bush's proposals. Bush's policy is nothing more than his father's failed* Battlestar Galactica *approach, which was floated around in the early '90s. It didn't make any sense then and certainly doesn't now, even though Bush is trying to resurrect a cold/space war with China as the adversary. Bush's policy ignores the more sensible Mars direct plan, proposed by Robert Zubrin.*

I don't discount the political aspect of both announcements, but it is too easy to react with cynicism or an anti-Republican attitude. And preparing for China as a potential future adversary is prudent. However, the current Bush initiative clearly calls for using the private sector to free NASA from its mundane activities, as I stated in my essay, and I don't recall the first Bush initiative doing so.

Regarding Zubrin's "Mars Direct" plan, I don't view it as more sensible to bypass the Moon and creation of an economically viable Earth-orbit infrastructure in order to quickly reach Mars. This is too similar to the "flags and footprints" Apollo missions that lost public and political support. I believe it is more sensible to take a building-blocks approach, where we first become proficient in our own Earth-Moon neighborhood and use that as a foundation for expanding to Mars and beyond.

5) *The Author also has a very optimistic or maybe even a trifle naive view of the part corporations can play in the role of space exploration.*

Yes, I am very optimistic, based on the historical accomplishments of corporations at creating new products and services based on new technology. I would ask the reader why he/she is pessimistic, and why space must be viewed differently from all other human endeavors that successfully combine private/public sector strengths. The government often leads, but it is business that determines how to make something profitable and cost less over time for all of us. At one time, government researchers were the major visitors to Antarctica. Now you or I can take a cruise there. At one time, only the wealthy could fly or sail across the ocean. Today, millions of people can afford such activities. Why? Because the private sector made it affordable and accessible.

The history of NASA's precursor, the National Advisory Committee for Aeronautics (NACA) provides the necessary insight. The development of a nascent commercial aviation industry was greatly supported by federally funded research and development.

Consider NACA Report 18, "Aerofoils and Aerofoil Structural Combinations," written by E.S. Gorrell and H.S. Martin in 1917. These weren't government efforts to own and operate a fleet of aircraft; this was research available to and used by the private sector for the creation of aircraft by the private sector, owned and operated by the private sector, and used by paying customers. The result was one of the largest sectors of our economy that is crucial to a broad array of activities. Why can't space happen the same way?

6) *I'd also recommend that the author read Wyn Wachhorst's book* The Dream of Spaceflight. *The author seems to have completely missed the perspective change; life altering, dreams of the dreamer; and exaltation of space flight. Wachhorst will help him with that.*

I write about space, talk about space, and lobby for space precisely because I am a dreamer who sees that space is integral to the human condition. My dream is far bigger than NASA programs. My dream involves all of us, not just astronauts. If this dream is to become a reality, if you or I or our children are to have a chance to go into space as anything other than government payroll astronauts, then there is only one way to accomplish this—the private sector.

SPACE: TEENAGERS AND THE FAR OUT   **113**

I want what this reader seems to want; the life-altering, perspective-changing experience of being in space and seeing the Earth from several hundred miles up. I want everybody who can afford an airplane ticket to ultimately be able to afford a flight to an orbiting hotel; and further into the future, space cruises to the Moon and beyond.

I hoped to convey my excitement about space by referring to "the Human-Space Connection™," with my comments in footnote 8, and with the following paragraph on pages 102–103:

"When we stop looking at space as a 'program' and start thinking of space as just another 'place,' we discover exciting new personal ways to relate to it. The first step is realizing space is the vast neighborhood in which we actually live. It isn't a distant and impenetrable domain. It is a mere 62 miles above Earth and thus a matter-of-fact continuation of our environment. You don't have to go very far to experience the darkness, the weightlessness, and the magnificence of space as you look down at the rotating Earth and watch your hometown go by."

If I failed to convey my sense of wonder about space, it is entirely my responsibility to make this clear in future essays.

DISCUSSION QUESTIONS

1) Should governments explore space when there are problems to solve at home? Why?

2) Should corporations and individuals be permitted to own property in space, like near-Earth asteroids and plots on the lunar surface? Why?

3) If nations that are current or potential adversaries (like North Korea, China, Iran) continue to develop their nuclear and/or space-launch capabilities, should the United States develop space weapons or rely on international treaties? Why?

4) Should all technology that has military potential be banned from use in space? Why?

5) What role should the United Nations play in space exploration, settlement, and development? Why?

Courtesy of The Venus Project
Designed by Jacque Fresco and Roxanne Meadows

■ Essay Eleven ■

DEBATE: SPACE AND OUR NEAR FUTURE

Jeff Krukin
www.JEFFKRUKIN.com

If you were to ask me, a space enthusiast, I think teenagers like you should jump into a hot debate likely to rage for the rest of your lives, and long into the lives of your yet-unborn children. Here are just some of the major issues as they relate to President George W. Bush's space plan, which he unveiled in January 2004. The transcript can be found at www.whitehouse.gov/news/releases/2004/01/20040114-3.html:

ENORMOUSLY EXPENSIVE, DANGEROUS, AND VEILED

Rebuttal: Several words deserve particular attention and will be considered individually.

A) "Enormously expensive": This must be considered within several contexts, beginning with the size of the federal budget. President Bush's Fiscal Year 2002 budget request for the U.S. government was $1.929 trillion, including mandatory and discretionary outlays.[1] Of that amount, $13.6 billion[2] was requested for NASA, which is just 0.7 percent of the budget. Does this qualify as an enormous amount?

Another context is the planned increase for NASA's budget. In his January 14, 2004, speech, the president called for "Congress to increase NASA's budget by roughly a billion dollars, spread out over the next five years."[3] Within the context of the federal budget, is this an enormous amount?

Yes, after the next five years NASA's budget will increase more, but it will always be a miniscule portion of the federal budget. The context of spending must also be seen within a historical context of how we've "done" space versus the future of how we must "do" space. If you only see future space activity through the lens of the past, then all you will see are massively expensive federal programs.

President Bush understands that the government cannot do this alone. Part of the White House document "A Renewed Spirit of Discovery" instructs NASA to "pursue commercial opportunities for providing transportation and other services supporting the International Space Station and exploration missions beyond low Earth orbit."[4] Involving the entire U.S. economy, rather than just the same few aerospace companies working on large government contracts, is the only way to reduce the cost of Bush's initiative.

B) "Dangerous": Any space undertaking is dangerous today. So were the first aircraft of yesterday; "By Oct. 14, 1911, the popularity of air meets, the barbaric demands of the crowds and the machismo of pilots worldwide had combined to bring the number of flying machine fatalities to an even 100."[5] As technology progresses and we gain experience, transportation becomes less dangerous. Why shouldn't this also happen in space in the same evolutionary manner? First, Earth-orbit travel improves, then Earth-Moon, and eventually Earth-Mars.

I realize there are experts who say it cannot be done. This is frequently the case, no matter the issue. Still, "in the beginner's mind there are many possibilities, but in the expert's mind there are few."[6] The Wright Brothers were beginners, and look what they did.

C) "Disguised as a noble effort to hunt for the 'origins of life'": Where's the disguise? The "Goals and Objectives" section of the document "A Renewed Spirit of Discovery" is crystal clear: "The fundamental goal of this vision is to advance U.S. scientific, security, and economic interests through a robust space exploration program."[7]

UNNECESSARY CONFLICT

Rebuttal: Again, let's consider specific words, as these words have great emotional weight.

A) "Create unnecessary conflict" and "weapons into space": How will conflict be created, and how is this conflict defined? Might the arms race extend into space? Yes. Is this the same as actual conflict? No. Does it guarantee conflict? No. In the twentieth century, the U.S.-Soviet arms race prevented direct and massive war between the two nations. It was called MAD, "Mutually Assured Destruction," and it worked because it created a balance of power where trust was lacking.

Do you believe that if the United States chooses not to place weapons in space that all our real and potential adversaries will automatically do the same? Isn't it possible another nation may use this to their advantage, putting weapons in a place where we gave up our defensive capabilities? These are important questions and deserve thoughtful consideration.

Are treaties the answer? Only when everybody abides by the treaties. If a potential adversary doesn't abide by them and has military capability in space before us, we will be at a tremendous disadvantage. We will have lost the edge.

I would prefer that space be free of weapons, but I don't see that happening as long as nations act primarily in their own interest. This is who we are at this point in our history. Wishing we didn't have to prepare for future conflict doesn't eliminate the possibility of conflict, not when some powerful nations aren't ruled by the same peaceful values as democratic nations.

B) "Nuclear power into space": What specifically is the argument against nuclear power in space? That it might be used as a weapon? If that's it, should we also ban all future power and propulsion technologies?[8] They will likely be useful as weapons, too. This attitude can be taken to an extreme, like banning all access to space because of its potential military use: no communication or weather satellites, no spacecraft, no exploration, no settlement.

If this attitude were successful in the past, think of all the dual-use (military and civilian) technology we might not have today. Ships, aircraft, and automobiles are all vital to our lives, and are valuable military tools. What would our society be like without them?

Is this simply an emotional antinuclear sentiment? What is the alternative that can't possibly be used in some destructive manner? If humanity is going to explore *and settle* the solar system, nuclear power is necessary for long journeys until we develop something even more powerful. All technology can be abused, but we are a technological species; there is no turning back.

To eliminate a useful technology or endeavor because it has military and peaceful uses is an easy choice, but is it a wise choice? Science, technology, and space are vital to our economy and way of life. What's required is thoughtful and balanced consideration, rather than simple avoidance altogether.

TAKES ADVANTAGE OF NEVER HAVING SIGNED THE 1979 MOON TREATY

Rebuttal:

A) "Land claims": Should the Moon be completely untouched by human hands? As it's quite large, isn't it possible to use it in the same varied ways that humans use Earth?: as a research station, mining facility,[9] college campus, shopping center, undeveloped park. Why should it be different?

I realize that some believe space must be protected from the ravages of humanity, because we will only exploit it. Many environmentalists want to protect space from being spoiled, just as they wish to keep Earth clean. This is commendable, but humans need resources to live. Where shall we get them? What would you rather keep unspoiled, Earth or small asteroids containing metal ores needed on Earth?

China and India are the two most populous nations in

the world, with a combined 2.3 billion people. Their economies are growing into twenty-first century powerhouses that will require vast resources. How will the governments of these nations provide for their citizens? Will Pearl Harbor be attacked again, only this time China destroys it with a nuclear strike before deploying its navy to seize the Spratley Islands[10] and other oil-rich parts of Southeast Asia? What will India do, situated between the oil wealth of the Middle East and Southeast Asia? Is conflict the only answer? Is conservation alone a realistic alternative? Must Earth be the sole source of energy and other resources?

B) "Military bases": If all nations will adhere to a treaty banning military force on the Moon, I'm all for it. A U.S. decision to keep its military Earth-bound can't be made in a vacuum, pardon the pun, without considering what other nations may do in space with their military. Space is the ultimate high ground, and many nations are developing their spaceflight capabilities.[11]

China is a potential adversary with ambitious plans for its spaceflight capability. Russia is a potential adversary when its economy gets stronger. Which government (not administration, this isn't about President Bush) and military do you believe is more likely to protect your way of life: American, Chinese, or Russian?

Many nations welcome the presence of U.S. military forces. U.S. military bases around the world have maintained peace in potentially volatile areas. Singapore welcomes U.S. Navy ships patrolling nearby. Global commerce depends on the sea, and our Navy keeps the sea lanes open. In a chaotic and competitive world, our military provides a needed deterrent. This may be necessary when Earth-orbit and Earth-Moon commerce become a reality.

If a military base is the first base on the Moon and leads to scientific, academic, and commercial bases, that's fine with me. As long as it belongs to us or a nation we can

MOVING ALONG: FAR AHEAD

trust with our lives. If there's a missile or laser platform in orbit, I would rather it belong to us than to North Korea, Iran, Russia, or China.

MONEY BETTER SPENT ON SOCIAL PROBLEMS

Rebuttal: On the face of it, this seems to make perfect sense. Just one question: When will there *never* be pressing social problems?

Do you know of any nation, any society, or any family that has solved all of its problems? How do you define "social problems" in a way that you know when they are all solved, and who does the defining? We will always have problems to solve. Such is the nature of humanity, a dynamic and ever-shifting force always changing itself and its social and political institutions.

This attitude would have all 6 billion human beings still living in Mesopotamia, the cradle of civilization, because nobody would have explored and settled elsewhere. It would be a very crowded and miserable place to live today.

Human beings explore in order to learn, to grow, to create a better life, to avoid stagnation. Do you stay home until you solve all your problems, or do you sometimes leave home because you must look elsewhere for solutions? Will you go to college, or will you solve your problems by staying home after graduating from high school?

NASA's Fiscal Year 2002 budget was $13.6 billion. In the same year, the federal budget for education, training, employment, and social services was $165.9 billion.[12] The budget for health, Medicare, and veterans benefits and services was $628.6 billion.[13] How will taking NASA's funding and spreading it across all our social problems magically solve them when the current expenditures haven't? This nation has the wealth to invest in its future while also continuously addressing its immediate needs.

As our economy expands into orbit and then to the Moon, new businesses and jobs will be created. Rather than simply have a government space program drawing from the U.S. Treasury, these businesses will generate revenue going into the

Treasury. Now you have more tax dollars to spend on social problems.

If we are to survive as a nation, we must invest wisely to continuously expand economic well-being for all our citizens. When a society becomes less able to sustain itself, individual rights are inevitably trampled upon as the powerful use all means to provide for themselves. When a society fails to seek knowledge and grow, it withers and dies. Without vision, there is little motivation to grow.

[NOTE: The four arguments against the Bush space initiative used in this essay come from a press release written by Bruce K. Gagnon, coordinator of the Global Network, and Dr. Michio Kaku, a theoretical physicist, which can be found at http://www.globalresearch.ca/articles/ GAG401A.html]. Having read this, I believe this organization's views are skewed by the belief that business profit is inherently bad. As I said earlier when speaking of technology, there is no room for balance in this view. Neither profit nor technology are inherently bad; it is how they are used that matters. Without the profit motive, we wouldn't live in a comfortable and progressive society. Communism failed, and socialism does include various forms of profitable capitalism.

A critical assessment of the Bush space initiative can be found at the Web site of the Global Network Against Weapons and Nuclear Power in Space: www.space4peace.org. Founded in 1992 to stop the nuclearization and weaponization of space, it has more than 170 affiliate groups all over the world.

NOTES

1 Office of Management and Budget. *FY2002 Budget of the United States Government* (Washington, D.C., 2001), 18.

2 *Ibid.*, 30.

3 "President Bush Announces New Vision for Space Exploration Program." www.whitehouse.gov/news/releases/ 2004/01/20040114-3.html.

4 The White House: www.whitehouse.gov/space/renewed_spirit.html.

5 Stanley W. Kandebo, *Aviation Week & Space Technology*, "The Wright Brothers and the Birth of an Industry," 30 December 2002, 37.

6 Zen master Suziki Roshi. I also like the noted author and scientist Arthur C. Clarke's statement, "When a distinguished but elderly scientist states that something is possible, he is almost certainly right. When he states that something is impossible, he is very probably wrong."

7 The White House: www.whitehouse.gov/space/renewed_spirit.html.

8 NASA Glenn Research Center: http://www.lerc.nasa.gov/ WWW/PAO/warp.htm.

9 What has space got to do with your way of life? "Gerald Kulcinski of the Fusion Technology Institute at the University of Wisconsin–Madison estimated the Moon's helium-3 would have a cash value of perhaps $4 billion a ton in terms of its energy equivalent in oil." Jim Wolf, "U.S. Eyes Space as Possible Battleground," 18 January 2004: www.globalsecurity.org/org/ news/2004/040118-space-battlegroud.htm.

10 "Rich fishing grounds and the potential for gas and oil deposits have caused this archipelago to be claimed in its entirety by China, Taiwan, and Vietnam, while portions are claimed by Malaysia and the Philippines. All five parties have occupied certain islands or reefs, and occasional clashes have occurred between Chinese and Vietnamese naval forces." www.inside-countryinfo.com/html/spratley_islands.html.

11 Global Security.org's World Space Guide: www.globalsecurity.org/space/world/index.html.

12 Office of Management and Budget, 16.

13 *Ibid.*, 16–17.

"Take me to your remote control."

■ Essay Twelve ■

WHY SHOULD WE SEND HUMANS TO MARS?

Thomas Gangale
Executive Director, OPS-Alaska

In September 2003, one of my professors in international relations asked: "Why do you want to send people to Mars? Is it not better to focus on robotics for now?" I thought to myself: It is *cheaper* to explore with robots but not necessarily *better*. Despite the advance of technology, there remain tasks that humans can better accomplish using machines on site rather than via remote presence.

In 1969, NASA presented a plan to the Nixon administration to land humans on Mars twelve years later. The report by President Richard M. Nixon's Space Task Group concluded: "NASA has the demonstrated organizational competence and technology base, by virtue of the Apollo success and other achievements, to carry out a successful program to land man on Mars within fifteen years."

Since that time, there have been no insurmountable barriers to landing humans on Mars—except societal will. With each robotic mission to Mars, with each new advance in technology, the technical problem of sending humans to Mars becomes easier. What once were "known unknowns" become "knowns," and "unknown unknowns" become "known unknowns." Once we know we don't know something, we can research the problem and master it.

This is not to say it would not be a difficult, dangerous, and expensive endeavor. It will be. However, at this point, we are far better prepared to send humans to Mars than we were to send humans to the Moon when President John F. Kennedy made the decision to do so in 1961. At the time Kennedy issued his stirring challenge to the nation, the United States had only fifteen minutes of experience in human spaceflight—none of it actually in orbit around the Earth—yet eight years later humans walked on the Moon.

In 1961, we had not sent a single successful robotic mission to the Moon—much less to any planet—yet eight years later humans walked on the Moon. In 1961, we had launch vehicles capable of putting only a couple of thousand pounds into orbit around the Earth—yet eight years later humans walked on the Moon.

In the thirty-five years that it has been feasible to launch a humans-to-Mars program, we have chosen not to. We will do so when the necessary social and political forces align, but when that will be is difficult to predict. It could happen tomorrow, or it might not happen for generations.

Perhaps the desire to go to Mars can be explained in part as a cultural after-image of Lowellian Mars.[1] Victorian civilization was convinced that it was on the verge of making "Contact." It was an age when the *New York Times* reported Nikola Tesla's plans to send radio waves to Mars and communicate with its inhabitants. As we better acquainted ourselves with Mars in the scientific sense in the twentieth century, there came, as H.G. Wells wrote, "the great disillusionment." We came to realize that in terms of sentient species, we are alone in the solar system.

Yet a faded echo of Lowellian Mars remains. We cling to the hope of a neighboring planet that harbors—if not canals and an advanced civilization—at least some primitive forms of life. If Mars contains even nanobacteria—or indisputable evidence of past life of the simplest forms—this will profoundly change our concept of our place in the universe. If there is—or was—another Genesis here in our solar system, then life must be common throughout the universe, and "Contact" with another civilization is therefore inevitable.

Do we need to send humans to Mars to discover this? No, not necessarily. It is possible that robotic missions to Mars could make such a startling discovery. But machines alone are not as capable as humans and machines working together on

126   MOVING ALONG: FAR AHEAD

site. So, if robots do not find life on Mars, the question remains open, even if just a crack. Eventually, we humans must go to Mars ourselves to definitively satisfy our curiosity.

As forbidding an environment as we have come to know Mars to be, it is nevertheless the most Earth-like planet in the solar system; the most readily accessible from Earth. And given sufficient technology and infrastructure, it will be able to support human life. It is true that Mars is a far cry from our own abundant, life-giving world. The photographs returned by the first robotic fly-by probes in the 1960s should have erased forever the previously held romantic mental images of Mars, but perhaps they have not erased them entirely.

Perhaps these are the true "ghosts of Mars," the spirits of our own past imaginings, and perhaps this is because we want to have neighbors on another world, because we do not want to be alone. Perhaps this is because, even if we cannot make "Contact" with the Other, the Alien, in our own solar system, we do not want to be confined to this Earth.

Is it worth spending tens of billions, possibly hundreds of billions, of dollars to send humans to Mars? In considering this prospective question, it is useful to ask a retrospective one: Was it worth it to send humans to the Moon?

There are certain indelible images of the age of photography: Battleship Row in Pearl Harbor on December 7, 1941, Abraham Zapruder's film of Dealey Plaza on November 22, 1963, the twin towers of the World Trade Center on September 11, 2001. These not only capture specific events but also define the specific locales and eras in which they occurred. But the images of Earth that we brought back from the Moon are timeless and universal, because they are the first images of *all* of us.

Ever since then, because of those images, we have looked at ourselves, each other, and Earth in a new way. The image of the full Earth brought back by the last crew to return from the Moon is an enduring icon of environmental responsibility and human unity. Was it worth going to the Moon to bring back even one of those photographs of Earth? I believe that it was.

The most important thing we discovered on the Moon was part of ourselves. In the few hours a few of us spent on the Moon from 1969 to 1972, we became better earthlings. As the poet Archibald MacLeish wrote, we were "riders on the Earth together." We realized we were our brother's keeper, and we remembered God had appointed us stewards of Earth.

And yet, more than a third of a century later, we must reflect on how pitifully less we have done with that revelation than we should have. It is high time we journeyed outward to that distant perspective, to see again how close we really are to each other, and to relearn those lessons that have faded with the passing of a generation. There are new lessons to be learned on Mars. There are new poems waiting for us.

If Mars is dead now, but was once alive, understanding how Mars died may give us a crucial understanding of how close we are coming to killing Earth. Also, just as no one could have foreseen the transformation of human consciousness that going to the Moon brought about, no one can predict the further transformative experiences of going to Mars. However, history suggests this will be the case.

How we go to Mars is as important as whether we go. In the twentieth century, a single nation went to the Moon on a Cold War double dare. In the twenty-first century, let it be a united Earth that goes to Mars. Going to Mars, then pushing outward to the stars, will be a parallel process with other human developments in a push-pull relationship.

Going to Mars together will go hand in hand with coming together here on Earth. Bringing life to Mars will go hand in hand with assuming responsibility for the competent stewardship of life on Earth. Bridging the gulf of space to meet and understand the Alien will go hand in hand with tearing down the obstacles of greed and prejudice that are the source of alienation on Earth.

The science fiction novelist Robert A. Heinlein wrote: "The Moon is a harsh mistress." All of the new worlds will be harsh. We will live close to the edge of extinction out there, and learning to survive on those other worlds will bring us closer to immortality. We will learn to depend on each other for our very lives as never before—Africans, Americans, Asians, Australians, Europeans, all of us. The New Frontier will be punctuated by tiny habitat modules, not sprawling with the wide-open spaces of the American Old West. We will live in enclosed places, in each other's faces.

All the pretentious barriers that we erect here on Earth will melt away in space. We will come to know each other—and ourselves—as we have never done before. We will push the outside of the envelope of what it means to be human. Living together so closely, so intimately, so inescapably, will tear down social and psychological walls we need not and dare not consider here on our comfortable, capacious, suburbanized, subdivided Earth. There will be new challenges to human dignity, privacy, individuality, intimacy, and polity. One wonders whether Kennedy grasped the full import of his own words: "We set sail on this new sea because there is new knowledge to be gained and new rights to be won."

I am an engineer, and I am studying to be a social scientist. I am supposed to be dispassionate and logical. But after long pondering my professor's question, I find I come up short. Exploration is always to some degree a leap of faith into the unknown; it touches the human heart, which cannot be weighed on a double-entry ledger of risk and profit. As many are the rationales that can be offered in favor of exploration, as many can be counter-posed. Faith cannot be explained or defended rationally. Bounded only by the ever-expanding limits of the possible, the greater the challenge, the greater the human appeal of the endeavor.

Our parents' generation went to the Moon. Now it is *our* time. Will we go to Mars? Will we let our children dance among the stars? Will we take the leap?

QUESTIONS TO PONDER

1) Suppose robotic space technology could be made better. Should we wait to send humans to Mars until we have exhausted everything we can find out through robotic missions? Consider the pros and cons.

2) President John F. Kennedy demonstrated a strong political will to send humans to the Moon in the 1960s. What might be the obstacles to the political will to send humans back to the Moon or to Mars today?
3) Why did people in the late 1800s and early 1900s think that an advanced civilization might exist on Mars?
4) How would contact with a more advanced space-faring civilization change our own world?
5) What might you discover about yourself living cooped up in a small habitat with several of your classmates for more than a year? (The only trips outside would be few and far between, requiring a labor-intensive process of donning a heavy spacesuit, and a return to the habitat before your air supply ran out.)

NOTE

1 In the late nineteenth century, many astronomers believed that Mars had an Earth-like environment that sustained life. Some astronomers, such as Percival Lowell, went so far as to suggest that Mars was the home of an advanced civilization.

REFERENCES

Croswell, Ken. *Magnificent Mars*. New York: Free Press, 2003.

Daley, Tad. "Our Mission on Mars." *The Futurist*, September–October 2003.

Hartmann, William K. *A Traveler's Guide to Mars: The Mysterious Landscapes of the Red Planet*. New York: Workman Publishing Company, 2003.

Kluger, Jeffrey. "Mission to Mars." *Time*, 26 January 2004.

Morton, Oliver. *Mapping Mars: Science, Imagination, and the Birth of a World*. New York: Picador USA, 2003.

Mars: The Red Planet Collection (DVD). Narrated by John Lithgow, Brentwood Communications, 2000.

WEB SITES

Martian Time
  www.martiana.org/mars
NASA's Mars Exploration Program
  http://mars.jpl.nasa.gov
Center for Mars Exploration
  http://cmex-www.arc.nasa.gov/

© The New Yorker Collection 2004 Wayne Bressler from cartoonbank.com.
All rights reserved.

■ Essay Thirteen ■

A MARS COLONY: CO96*

John A. Blackwell, Phil Gyford,
Glenn Hough, Alexandra Montgomery,
and Dana Wilkerson-Wyche

Mars offers an opportunity for a fresh start for humanity.

OVERVIEW

Mars Colony 096 is both an experiment and an evolutionary phase change for humanity. While others have lived far from Earth for many years and some have been born alien to our solar mother, we are the first to purposefully redefine the reason for our being and to choose and design an existence that leads us to our common purpose.

We are unfettered from earthly demands, by our charter, by our distance, and most important, by our choice. We choose to live free of need by giving, and free of fear by loving, and with these human energy sinks removed, we dedicate ourselves to the generation of harmony throughout the universe.

Ours is an experiment because our test may fail, leaving Earth and its colonies as the remaining example for human survival. It is an evolutionary phase change because having embarked on our quest we can never return. Should we survive and even thrive as we expect to, every single encounter, within our colony and without, will support and be guided by an abundant view of an infinite universe.

We harbor no condescension for our forebears but thank and honor them for enabling the potential we now hold. It is their advanced technology that allows this land of plenty. Micro-robots operating at a molecular level construct our foods, tools, and buildings. They generate clean air and water and provide a zero-waste environment. Our communications systems make physical presence essentially indistinguishable from virtual presence. We have harnessed the power of medicine, and evolved to include both ancient and modern technologies, which allows us to lead a healthy lifestyle and live for 150 years.

We learned from Earth that governance systems built on power inevitably fail because there is always another greater power. So, we have chosen to base our governance on self and abhor all forms of violence and its use to exert influence. We believe in democracy and have chosen to combine elements from Earth's governments that support our Code of Ethics.

Our representatives are chosen by lottery and their purpose is to help the community understand our opportunities for using our local, human, and extra-colonial resources. The community decides directly which projects will be done. Because every member of the colony contributes time to these efforts, we are continually working on those things that bring us pleasure and closer to our common vision.

We recognize that excellence can be achieved through both competitiveness and the desire to serve. We have chosen service as our driver, and therefore all of our needs and wants are met through our giving to the community. We do not have physical money but trade value for value in anything that our community has chosen not to provide equally to all. In this way, individuals may engage in any activity that does not infringe on the rights or common access to resources of others for their own profit or pleasure.

Most of our time is our own, as provisioning our community is done easily once accumulation is eliminated, and a diverse array of services and goods are provided and bartered for. No one may accumulate more than ten times the credits given to every member of the community for their basic service. That value simply disappears from their accounts and becomes part of the communal larder.

We expect to encounter many problems, both within our own colony and with the rest of the universe as we follow through on our commitment to engage and explore. We will use our wits and be guided by our belief that peace, love, and harmony are not just the forgotten mantra of a time more than one hundred years past but universal truths that could be seen then and lived now.

THE GOVERNMENT

The process of government is designed to offer a balance between direct democracy and a representative system. It is hoped that both the multiple-choice ballots and the random appointment of representatives from the entire population will make the electorate more involved in the entire process and discourage an "us versus them" attitude to government.

The aim is to create a consensual, rather than adversarial form of government within practical limits. Both the electorate and the single legislative body, the Senate, must agree on a bill for it to become law. Other features of the system are an extensive committee structure and a president whose role is mostly ceremonial.

STRUCTURE

The Senate consists of one hundred people, each serving a term of four years. Every year twenty-five are replaced by new representatives chosen randomly from the population. Appointments occur a couple of months before the end of the year, giving new senators time for training and job shadowing before beginning their term.

The colony has a president as head of state although he/she holds no direct power, acting as a ceremonial figurehead and the public face of the community. The president is drawn from the general public, and anyone can put him or herself forward for the position. However, candidates must have an extremely high level of public service and commitment to the community. Those candidates who meet the highest standards are entered into a pool from which the new president is randomly chosen. The terms of office are four years.

THE COMMITTEES

There are many government committees, and every representative is expected to serve on at least one. Some of these are permanent committees, each overseeing the functions of an individual department, and many are temporary and devoted to researching a single bill. It is expected that senators serve on committees relevant to their experiences.

The committees have responsibility for allocating public resources. This system is meant to guarantee that resources are fairly dispersed and used so that they are faithful to the agreed-upon goals of the community. There are eleven standing committees that are understood to be necessary to the survival, culture, and standards of life desired in the colony.

At this time these eleven committees are meant to be permanent, but since it is the nature of the committee system to be fluid, this can easily be changed. Temporary committees may be formed specifically around the introduction of a bill, policy implementation, or other community projects as they occur outside of the realm of the original eleven committees.

"A basic question that any society should confront is how to govern for the benefit of all its members." [1]

Senate members voluntarily sign up for the committees of their choice upon allotment and as vacancies occur (when a senatorial term is up every four years, so is committee service). It is likely that one senator will serve on at least two committees. One would feel inclined to choose a committee in which they have competency, interest, and some degree of expertise or training.

The eleven standing committees:
1) **Arts:** Sponsors art, theater, music, creative expression.
2) **Community:** Child care, elder care, community activities, spiritual support.
3) **Education:** Sets basic curriculum, allocates funds.
4) **Emergency (the cabinet):** Manages natural disasters, civil unrest, medical emergencies.
5) **Health:** Reproductive issues, hospitals, disease control, preventive care.
6) **Holism:** Systems-oriented, long-term view of the whole colony.

7) **Judicial:** Mediation, arbitration, rehabilitation, Peace Corps.
8) **Structures:** Buildings, structures, construction and maintenance.
9) **Technology:** Explores new technologies.
10) **Transportation:** Planning, consulting, regulation of public transportation and infrastructure.
11) **Utilities:** Energy, water, food, air, shelter.

FEEDBACK STRUCTURE

An egalitarian method of monitoring daily operations of the colony is enabled by the technology available in 2080. To make information available to the public and to keep up with public opinion, an automated feedback structure will be devoted to preventing social problems. A computerized system would allow people to submit scores reflecting satisfaction or dissatisfaction with ideas, projects, or circumstances in the colony.

The system does not intend to privilege empirically gathered information to the human element of public opinion; instead it is intended to account for easily measured and easily obtained expressions of social concern. Each committee monitors daily activity in areas relevant to its field. The Holism committee is constantly interpreting the data to ascertain that the community is consistent with the values and goals of direct democracy.

CRISIS DECISION-MAKING

The decision-making process (as indicated in Passage of a Bill, below) can be a lengthy and repetitive process that, while oriented toward gaining a wide consensus on an issue, is not suitable for occasions when speedy decisions are required. In times of crisis, any member of the Senate can propose that the Cabinet gain control over decisions related to the crisis. Seventy-five percent of the senators must approve the temporary measure.

While the Cabinet has control, any senator may propose that the crisis is over and normal processes should resume, again with 75 percent approval. No decision made by the Cabinet should be permanent; it may be overturned by a bill issued in normal manner if the orignal bill does not expire with the end of the crisis.

PASSAGE OF A BILL

All votes are in a multiple-choice format, that is, giving the voter more options than merely approving or vetoing a proposal. This is intended to elicit more thoughtful, subtle, and considered responses. In addition to a straight-out Yes or No, there are options like "No, I disagree with this formulation but do not disagree with the principle."

"The concept of citizenship began to take place as individuals and their associations organized to contest and balance the authority of the state, of the nobility and the clergy."[2]

Whenever a stage involves a vote, there are usually five potential outcomes: *Y1, Y2, Y3, N1, N2. Y1*: Yes, full support. *Y2*: Yes, but with reservations. Yes to principle, but work needs to be done on wording, context, procedure, etc. *Y3*: Yes, passive support. No real opinion or interest in this measure. *N1*: No at present, but willing to talk about changes and/or compromises. *N2*: No, absolutely not.

Where a bill goes from here depends on the outcome of the vote. A 66 percent Y1 vote will pass and enact a bill the same day. A 66 percent N2 vote will reject a bill utterly and cause it to be dropped completely from consideration, where it can only be resurrected if it goes through the process from scratch.

If a bill has not even a simple majority, in any of the categories, then more work needs to be done on it. The bill has failed the most basic tenant of actively trying to form consensus.

A bill may be initiated either by a government department (headed by a committee) or by a nongovernmental entity—a person, a company, or an organization. Anyone outside the government may propose a bill, which is added to an electronic list of currently proposed bills. Anyone may add their digital signature to support a bill and when/if a certain percentage of the population has signed, it moves to the next stage.

There is a time limit after which bills without the required support are retired and must be initiated from scratch. It should be illegal for companies to coerce their employees to support a bill proposed by the company, and there should also be restrictions on the advertising of proposed bills (if it is allowed at all).

THE JUDICIARY

The judicial reviews will be in an open forum. Panels formed from the judicial committee conduct the reviews. All citizens are encouraged to attend. The purpose of the judicial review will be to interpret the values of the community and determine if a value was violated.

Punishments are by nonviolent methods. The highest degree of punishment will be returning the individual to Earth for adjudication, and it will only be levied for violent offenders. The panel in open forum will review all other offenses; the people will determine the dispositions. The panel will determine the method of rehabilitation.

REHABILITATION AND MEDIATION PROCEDURES

Rehabilitation and mediation is designed to resolve conflict and behavioral problems within the colony: citizen to citizen, citizen to public authorities, and among the different organizations of government.

The method of rehabilitation varies depending on circumstances and violation. Rehabilitation consists of education, sensitivity training, etc. Rehabilitation will be used in disputes between citizens and public authorities and organizations of the government. A panel of individuals from the judicial committee will mediate to arbitrate violations and conflicts between citizens.

JUDICIAL POWERS

The judicial power of the colony shall be vested in three panels, and each panel will allocate rehabilitation with accordance to the global Code of Ethics and Bill of Rights. The Cabinet may ordain and establish legal procedures for the panels to follow. To maintain harmony, the Cabinet will ensure quality living by providing the essentials for general welfare, freedom, and a reasonable expectation of privacy for all citizens.

PEACE CORPS

"The most desirable form of political leadership ... no death penalty, no torture, no war, no terrorism, and with military and police forces are converted into unarmed specialists in nonviolent conflict resolution, the deglorification of violence and the appropriate redirection of military resources."[3]

Mission: To provide the citizens with a sense of personal security; to reduce or control violations of the code; and to establish a peace corps/citizen communications network through which mutual problems may be discussed and resolved.

The Peace Corps was designed by the judicial committee to maintain interpersonal interactions with the citizens of the colony. At the age of eighteen, citizens may volunteer for the Peace Corps. Each volunteer will attend a peace control academy, which will focus on methods for preserving peace and harmony. After completing the requirements of the academy, the citizens will be required to work a minimum of four hours per quarter. Their duties will be compatible with the Peace Corps mission. The four hours will be assessed to their twenty-hour workweek.

THE ECONOMY

Our colony is a land of plenty. For the required twenty hours of colony work per week, a citizen's food, lodging, and other basic amenities are lavishly met. A person really doesn't need to work any more if he or she doesn't want to. If someone does work extra, on colony projects, then that work is credited to their personal account. If a person contracts out with someone else, that person negotiates the amount of credit for the amount of work and then transfers the credits directly.

"Because the accumulation of human capital can replace some forms of exhaustible resources, human development should be seen as a major contribution to sustainability."[4]

There is no physical money in this system. Everything is done electronically. There is also a cut-off point in the system. An individual's personal wealth may not be more than ten times the base unit of twenty credits.

If people wish to start a business, they can petition for starting funds or materials. It will be decided if this assistance is given in the form of a grant, if the individual must work extra for this, or if the colony will become a silent partner and receive a percentage. If an Earth-based company decides that it wishes to be in the colony's market, it must find local owner-ship and control no more than 49 percent of the stock.

We envision that there is nothing like "retirement" as it is practiced on Earth. As people go through life, their "job" changes but the value does not. The entire system is geared toward finding value in each person and whatever it might be that they can contribute.

TASK ALLOCATION

Various Senate committees handle task allocation. The com-mittee of structures handles the physical buildings of the colony. We envision that there is a continuing computer-gen-erated pool of available project managers from whom all the committees can draw. Once a project manager is chosen, the committee sets the parameters of what needs to be accom-plished and then steps back into a supervisory position.

The project managers then use the same system to pull people for the task. This pool would include specialists they normally work with, people whose skills are just needed for this project, and newcomers who are changing positions or trying different tasks.

SOCIAL LIFE

Civic participation and leadership are indivisible aspects of everyday social life. This system requires that all citizens be prepared to take on leadership responsibilities at various points and in multiple contexts. Citizenship is understood as a respon-sibility to ensure the well-being of the whole of society, which in turn guarantees individual comfort.

CITIZENSHIP

All citizens are expected to work twenty hours (out of so many waking hours) a week toward the ensured subsistence of the colony as a whole. Participation in these activities is considered the most basic civic duty. All people contribute in some way, often on projects coordinated by one or two individuals and always in areas in which they have competence and/or interest. The activities range from beekeeping to technology.

"At its simplest, communitarianism is a movement based on an effort to balance individual rights with community responsibilities."[5]

Another way in which a citizen may be expected to participate is being allotted a position in the government. This would involve committee work, representation of the citizenry and/or colony, and legislative or other policy-related duties. While all citizens are encouraged to participate in assembly in general, being given a seat in the Senate for a four-year term would exempt someone from subsistence-oriented citizenship responsibilities.

Thus, the culture of the colony promotes public service and perceives it as vital to the endurance of the community (just as food production, technical maintenance, and habitat upkeep ensure survival, so do citizen participation and leadership).

Another component of citizen participation in this colony refers to the contribution of expertise and knowledge by the population to processes of governance. Besides the cultural emphasis on lifelong learning within the colony, high levels of education, literacy, and specialized training make the community a rich resource of information.

Senators are expected to use the wisdom, knowledge, common sense, proficiency, and experience of the populace. This component of the design tries to minimize wasted government resources by tapping easily available sources (people), and tries to maintain a sense of connectivity and harmony between the processes of the government and the governed.

LEADERSHIP

"For the majority of people, solutions for poverty alleviation, the security of productive employment and livelihoods, and the restoration of the integrity of the environment are rooted in private or community-based initiatives."[6]

Leadership is important to each of these three modes of citizen participation. Subsistence-related civic duties may be perceived as having rotating positions of leadership depending on tasks, number of people involved, length of project, etc. As members of the Senate, citizens could be elected to positions of leadership among the committees or within the cabinet. Positions of leadership are temporary, flexible, and conducive to electing dual leaderships.

As consultants to policy and legislation, the community as a whole is responsive to leadership. Being asked to contribute to the processes of governance is considered a rewarding opportunity to serve and emphasizes the concepts of Sadvipra (achieving progressive change for human development), servant leadership, and the ideals conveyed by the community values.

EDUCATION

We envision that the destructuring of the educational system shall open the pathways that great minds of the past have taken, to new generations.

Education is an extremely personal thing. Since no one must work more than twenty hours per week, educating children falls to the parents and then to the community. It is our hope that all children flourish and meet their potential. The system encourages the curiosity of youth, and this translates into a driving passion to know different things.

Once children have been taught the basics, which include survival training for Mars, they are gently guided into experiencing many different things before they have to decide on the direction for their life.

We envision that there will be groups throughout the colony forming around subjects of interest; breaking the learning process into collaborative team-learning, guided either by the group or a specialist. We envision that these groups will have members from many different age groups for some subjects and roughly the same age group for basic required knowledge.

SUMMARY
In our colony, design, vision, values, sociopolitical structure, the economic system, and infrastructure are defined so that the combined elements enable and support a people who seek to live in harmony with the universe. No one can know what kinds of challenges, threats, opportunities, or wonders may yet be encountered, but as each is faced, our actions will consider—as best we can—the impact on the seventh generation of our own kind—and all others.

APPENDIX: DESIGN PRINCIPLES
The following criteria provide a framework for making design decisions in such a system:

1) *The necessity of a common purpose or vision.* A new and compelling vision is needed for fundamental change to take place. A future significantly different from the competitive, confrontational, and deadline-driven existence that capitalist consumerism requires of many humans can be changed by seeking a purpose that is antithetical to such behaviors. Harmony is our guiding purpose.

2) *The necessity of structures that support stated values.* Values are at the root of choice and decision-making. Economic, political, and other structures must be consistent with our core values. As an example, we believe that competition breeds disingenuous behavior and therefore government representatives are chosen by lot. This eliminates the need for individuals to campaign and therefore shade the truth about themselves or others in any way.

3) *The desire for sociopolitical decisions to approach consensus.* While absolute consensus is not feasible as scale increases, systems can be designed to protect the interests and rights of citizens as individuals and as interest groups.

The intention is to remove the competition between interests and replace it with mechanisms that are inclined to encourage dialogue rather than debate. Minority issues must be highlighted and specifically considered in repetitive passes of the legislative process.

4) *The necessity of a multiple peer valuation system.* Capitalism is bottom-line oriented. Some progress has been made toward addressing (at least in words) environmental and social values along with profit as a value, but more can be done. A three-pronged approach is taken to achieve this end. First, extreme conventional wealth is mitigated against, and second, valuing each individual for that which they would be valued is promoted. Third, exercising conservatism with respect to all resources and systems, particularly those we do not understand, insulates us from the negative effects of progress as a good in itself.

5) *The desire to share in the development and support of the infrastructure.* People care most about what they invest their energies in. This criterion seeks to allow every individual the widest possible opportunity to contribute to the community through service. Ideally what one might think of as drudgery, another considers a pleasure. All citizens are required to contribute, but the system is biased to accepting that contribution on their terms with respect to timing and type of activity.

6) *The necessity of privacy in an essentially communal environment.* Spaces to be used in common are preferred over private property ownership. In particular, access to the most desirable spaces should be equally available to everyone. Individuals and families of all sorts require privacy to function, and this must be supported by the design.

7) *The necessity to abhor violence.* Violence begets violence. Conflict resolution methodologies other than violence must be used to settle disputes between individuals, the state and individuals, within the state, and between the state and outside aggressors when possible.

8) *The necessity to consider future generations.* No one can know what kinds of challenges, threats, opportunities, or wonders may be encountered, but as each is faced, our actions must consider as best we can the impact on the seventh generation of our own kind and all others.

Germane is our new **Bill of Rights**:

1) Citizens have a right to worship and exercise religious practices of their choice.
2) Citizens have a right to privacy in their living quarters and personal effects.
3) Citizens have a right to be heard by a judicial panel when facing disciplinary actions for or against their well-being.
4) Citizens have a right to express their opinion on public and personal matters.
5) Citizens have a right to take part in the political process, including but not limited to the decision-making process.
6) The disciplinary process will not subject an accused citizen to any form of physical abuse.
7) The judicial committee has the power to enforce the global Code of Ethics and values of the colony.
8) The judicial committee exercises the right to extract citizenship and order an accused to return to Earth when he or she exhibits any form of violence.
9) The judicial committee has a right, when necessary, to examine individual contributions to ensure fairness and maintain harmony.

* A longer version of this essay can be obtained from one of its coauthors, Glenn Hough (gally_angel@yahoo.com).

NOTES

1 Los Horcones, "Personalized Government: A Governmental System Based on Behavior Analysis," *Behavior Analysis and Social Action*, vol. 7, nos. 1 & 2, 42–47.

2 Robert Cassiani, "Financing Civil Society for a Global Responsibility," *Futures*, vol. 27, no. 2, 215–222.
3 Glenn Paige, "Nonviolence and Future Forms of Political Leadership," in *The Future of Politics*, edited by William Page (London: Frances Pinter, 1983), 158–170.
4 Francisco R Sagasti, "Knowledge and Development in a Fractured Global Order," *Futures*, vol. 27, no. 6, 591–610.
5 Allan Winkler, "Communitarianism," *Utne Reader*, November/December 1994, 105–8.
6 Cassiani, 215–222.

REFERENCES

Bahm, A.J. *Computocracy*. Albuquerque, N.M.: University of New Mexico Press, 1985.

Bell, Wendell. "World Order, Human Values and the Future." *Futures Research Quarterly*, Spring 1996, 9–24.

Cassiani, Robert. "Financing Civil Society for a Global Responsibility." *Futures*. Vol. 27, No. 2, 215–222.

Henderson, Hazel. "New Markets and New Commons." *Futures*. Vol. 27, No. 7, 113–124.

Lewellen, Ted. *Political Anthropology*. Westport, Conn.: Bergin and Garvey, 1983.

Los Horcones. "Personalized Government: A Governmental System Based on Behavior Analysis." *Behavior Analysis and Social Action*. Vol. 7, Nos. 1 & 2, 42–47.

Paige, Glenn. "Nonviolence and Future Forms of Political Leadership." In *The Future of Politics*, edited by William Page. London: Frances Pinter, 1983.

Sagasti, Francisco R. "Knowledge and Development in a Fractured Global Order." *Futures*. Vol. 27, No. 6, 591–610.

Slaton, Christa Daryl. "Quantum Theory and Political Theory." *Quantum Politics*. Theodore L. Becker, ed. Westport, Conn.: Praeger Publishers, 1991, 41–63.

Winkler, Allan. "Communitarianis." *Utne Reader*. November/December 1994, 105–8.

Epilogue

ON USING FUTURISTICS

*The important thing
is to not stop questioning.*
—Albert Einstein*

Questions, of course, remain, far more than the number of any answers possible from the fifty-eight essays in this series. And that is as it should be. We are far more interested in ratcheting up the quality of what it is that puzzles us than we are in dwelling on a new answer or two. Futuristics thrives on the generation of ever better questions, as these—rather than (always tentative) answers—can make a major contribution to decision-making, our primary concern.

We have come quite a way in improving the art of choosing. (In 1842, England issued a law decreeing that decisions made by "Rock, Paper, Scissors" were binding).[1] Advances in the related art of forecasting—as illustrated across the four books in this series—can aid further gains, as they help us better prepare to meet the unknown, both what we can imagine and that awesome other we cannot.

Much to the credit of practitioners, futuristics itself is steadily improving. Today, unlike decades ago, most forecasters decline to use the terms *will* or *will not*, recognizing that they imply far more certitude than is defensible. Today, they scan widely and seek clues in even seemingly outrageous material (cyberpunk literature, hip-hop lyrics, reality TV incidents, etc.). Today, they systematically and publicly critique forecasts—their own forecasts and those of others—to learn from error. Today, thanks to the Internet, they operate in a world-wide community and generously share insights, ideas, and emotions much as if enrolled in an Invisible College. With two outstanding master's degree programs beckoning to college graduates (the University of Houston-Clear Lake in Texas and the University of Hawaii at Manoa), futuristics would seem on the verge of an overdue renaissance.

Consistent with this notion of cultivating questions, and improving the art of forecasting and decision-making, the closing essay in this series hones in on an especially fine question: How can we finally become really and truly human? Insights taken from our history as a species are combined with a sense of glowing possibilities to suggest that we had best soon get on with it ... with cocreating a future that honors us all.
—Editor

* As quoted in an advertisement, placed by the Hyperion Corporation, *Wall Street Journal*, 23 March 2004, B-3.

NOTE

1 From an advertisement, placed by the Hyperion Corporation, *Wall Street Journal*, 23 March 2004, B-3.

■ Essay Fourteen ■

THE FUTURE OF US: HOW CAN WE FINALLY BECOME REALLY AND TRULY HUMAN?*

Melvin Konner, Ph.D.
Samuel Candler Dobbs
Professor of Anthropology, Emory University

WHERE WE HUMANS CAME FROM

There's good news and bad news about our species as we face the future. The bad news is the problems are huge. The good news is our species has solved worse problems. We are not just the ape that walks on two legs; we are also by far the best and most successful ape. We wouldn't be here, and we wouldn't be facing the problems we face except for our great success.

The Bible says, "Be fruitful and multiply." Evolution dictates something similar, and that is what we have done. We have dominion over the birds of the air and the fish in the sea and over every living thing that crawls upon the earth. In fact, we have dominion over every living thing except ourselves.

What do I mean by this? Consider our history. Our story began some 7 million years ago as a weak small ape that over thousands of generations, started to walk upright. It took several million more years before we began to expand our brain or to make the tools that turned around and helped make *us*. We created culture, and it created us. The Bible recognized that this was what made us different and what in time would let us rule the earth. What it didn't say, and what even we have trouble seeing, is that this in itself would prove our gravest threat.

Not just for a few million years but for a thousand times longer, the imperative shaping us was that of reproduction. And oh, did we do it. We don't look so great compared to bacteria or insects—there are a lot more of them, however you measure it.

But compared to other mammals, we are incredibly successful.

Most of our runaway success in reproducing came in the last few thousand years—in evolution, the blink of an eye. Yet it led not just to the evolution of *Homo sapiens*, but to a series of social and cultural transformations after our species came into being. Our first big advance was a spiritual one about forty thousand years ago, when our ancestors began to create great art on the walls of their caves. This put them in touch with some of the highest ideals human beings can aspire to, both religiously and artistically.

Then about ten thousand years ago, in several areas of the planet, our ancestors began to plant seeds and herd animals for a living. Why, after millions of years of hunting and gathering? Population pressure may have come first, forcing our ancestors to use knowledge they already had, or the idea of farming may just have taken hold.

Either way, the consequences were huge. Populations grew, and the new need to defend a piece of land planted by the sweat of our brows intensified our tendency toward war. Towns emerged, then cities—the hearts of empires. They were the products of conquest and taxation, the control of dominant people surrounding weaker ones. There were four kinds of control: control of water for irrigation, control of land and its produce, control of labor, and control of hearts and minds.

The first three came from armed force and the last from organized religion. Every ancient empire was ruled by a three-part elite consisting of armed nobility, priests, and a merchant class. Together they were irresistible to the common man or woman. Also crucial was the threat empires posed to each other. If you were a farmer on the edge of an empire, you might be brutally taxed and even enslaved, but at least you were alive. If the empire next door sent its army into your village, all bets were off.

So in a sense, what we call civilization was a protection racket. Violence was its hallmark, not poetry or sculpture. They were beautiful, of course, but incidental. It was once said of capitalism that it emerged from the mud with blood oozing from every pore. This may not be true of capitalism, but it certainly was true of ancient civilization: the mud of irrigated soil soaked with the blood of conquest.

FROM THE PAST TO THE FUTURE

So why include these ancient stories in an essay about the future? Because basically, not much has changed. Our world is still based on that ancient one. Bigger empires than ever struggle over resources—some the same, like land, food, and water; some new, like oil and information. Unbelievably powerful weapons have replaced swords and chariots, but the principle is the same, and our lives are still precarious, due to the threat of nuclear bombs and other weapons of mass destruction. The genocides against victims like the Jews of Europe, the Kurds of Iraq, and the Tutsi of Rwanda are another approach to mass extermination familiar from the ancient world but now much more efficient. And the attacks of September 11, 2001, show us how the very technology we are so proud of can be turned against us without mercy.

Population experts say that the human species will increase to "only" about 10 billion. Some say that there will be enough food, so all will be well. This is silly. First, we are degrading the environment so fast that the earth's future food supply may be lower, not higher. Production of wheat and rice is already declining. Second, wars in the near future will not be over food but over oil and fresh water. People as different as United Nations Secretary General Kofi Annan, former Soviet Premier Mikhail Gorbachev, and former U.S. Secretary of State Madeleine K. Albright agree that this new century, just born, could be overwhelmed by wars over water. Add to this global warming, the destruction of the ozone layer, deforestation, mass extinction, and other ominous environmental trends.

Finally, as September 11 taught us, our lifestyle inspires jealousy and hatred. At a minimum, it makes people throughout the world want it. But if the whole world got even a minimal American lifestyle—a small car, a little apartment with the usual appliances, a TV, a DVD player, and a computer—the planet's destruction would accelerate. It simply cannot support 10 billion people with that lifestyle, and no one with any knowledge thinks it can.

A TIPPING POINT

We seem to be boxed in, but there is another way to look at it. We are at a turning point or what some call a tipping point. We can follow a path to disaster, which is where we may be headed, or we can do a sort of historical judo flip that uses the weight of the disaster against itself. Humans unite against terrible threats and this one is bad enough to unify our sadly divided species. *But this will only happen if we understand the threat.* So knowledge is the first step but only the first. The second is will.

The gap between knowledge and will is shown by the fact that we know we are using up fossil fuels and warming the earth, but we keep buying more and more SUVs. These gas-guzzlers worsen our problems every day. Compared to the amount of money we put into buying and protecting foreign oil, our investment in research to find and use other energy sources—sunlight, wind, tides, ethanol, water, cold fusion, and many others—is minimal.

So one positive action needed is to wake up our government. This is most likely to be done by the same people who woke up previous governments about segregation and the Vietnam War: people your age. The young have vision and energy, they are open-minded, and they are neither committed to the status quo nor complacent about its dangers. You are the key to the tipping point process. Because of the bulge of births sometimes called the echo-boom—the baby boomers' babies—there are more people in their late teens and twenties every year. For the near future, the world depends on you.

IMAGINING A NEW WORLD

What should you do? Well, you begin by shaping the

THE FUTURE OF US　153

consciousness of the older generation; what happened in the 1960s proves that it can be done. And of course, you have to imagine a new world. Fortunately, that is not difficult.

Imagine if in the wake of September 11, besides everything else he did, President George W. Bush had announced a $100 billion program to end the United States' dependency on oil through conservation and energy innovation.

Imagine if the president said today, "My kids and nieces and nephews convinced me that I should sign an international treaty to end global warming."

Imagine if there were a $1 million prize offered every year for the best new idea on how to take the salt out of seawater.

Imagine if as much money were put into diplomacy, exchange programs, the Voice of America, the Peace Corps, and supervising elections in developing countries as is put into the U.S. Air Force, Army, and Navy. Or half as much.

Imagine if, besides the United Nations, there was a parallel international organization consisting only of democracies—an exclusive club that others would aspire to and *change* to be admitted to; each new member another beacon to the world.

Imagine if international agreements on the environment were not prevented by religious fundamentalists from mentioning family planning and population control.

Imagine if women throughout the world had equal rights with men and—as other liberated women have done—inspired their societies, protected the health of their families, educated their children, and reduced their family size.

Imagine if the Kyoto Protocol on climate change, the nuclear non-proliferation pact, the World Court, and the worldwide battle against AIDS became the models for scores, then hundreds of agreements that would move the world inch by inch toward international governance.

Imagine an effective ban against dictatorship, racism, oppression of women, torture, terrorism, and war.

BUILDING ON PAST SUCCESSES

Why should anyone with half a brain think that this is possible?

154 MOVING ALONG: FAR AHEAD

Because our species' successes in the past show we can do big things that once seemed undoable.

Slavery was once accepted throughout the world, then only in some countries, then it was banned. Once even the Bible took it for granted. It still exists, but it is on the run, reduced to pockets of renegade lawbreakers in very backward countries or in the underworld of advanced ones. With will and luck, your generation can erase it from the planet.

Democracy was once just a cool experiment. In 1776, it was pretty much a joke told about a bunch of dreamy hicks in a far-off wilderness who were dumb enough to think they could become more than a colony. Ha-ha. Now the Europeans who laughed have followed the American lead and are helping to spread the democratic message across the world. Democracy is built not over years but generations, yet it is already widespread and one day it will be everywhere.

Communism seemed for more than a century to be a possible alternative to capitalism. No longer. It collapsed of its own dead weight because it defied human nature, which becomes expansive and energized when the spirit of entrepreneurship—a sort of controlled greed—can function freely. That way, people who work for themselves and their families move the world forward. The world now agrees on this, and Communism is dead.

We've succeeded on other fronts, too. Smallpox was erased from the world by a huge international team, and polio will soon go the same way. In 1998, fewer babies were born than in the previous year, probably for the first time in human history. Population growth is slowing. It is still a major threat, but when you are thinking about retiring, it will peak and slowly decline. Not soon enough, but in the big picture, soon.

An international agreement banning CFCs—chlorofluoro-carbons, which still threaten the ozone layer that shields us from harsh rays of the Sun—included the United States, and it has stuck. No more CFCs are being made and that represents the first successful planetwide environmental pact. As for weapons, mustard gas was used in World War I, but the results were so terrible that—although Hitler and Saddam Hussein used them against civilians—gas attacks have not been used in a major war since.

WHAT CAN WE DO?

All these experiences tell us that our species can leave bad things behind and pick up new and good ones. But, you say, you are only one small person. What can you do? Talk to your minister, priest, rabbi, imam, or other religious leaders. Ask your teachers hard questions. Talk to your parents, brothers, and sisters. Write and call and e-mail your politicians again and again. Even if you can't vote yet you can let them know your opinions. Light your corner and help others light theirs. If you don't see a movement, get together with a few friends and talk about it until you see a path to change. The anthropologist Margaret Mead once said: "Never doubt that a small group of thoughtful, committed citizens can change the world. Indeed, it is the only thing that ever has." Often that small group has begun with one person.

It isn't a sure thing, not by a long shot. It's a tipping point, and that means it can go either way. Which side will you be on? The side of complacency and ladder-climbing and piling up stuff? Or the side of world-changing, bold imagination, and action?

It's a race against time. Within fifty years—easily within your lifetime—we will add four billion people. More than 200,000 a day, 80 million a year, a country nearly the size of India every decade. Four billion new mouths to feed, four billion thirsty throats, four billion dreams of a good life.

But it's worse than that. The poorest of the species will grow much faster than we will, and it will be very young. So, the new 4 billion of us will include a lot of poor young men. Will they have jobs so that they have some hope of living their dreams? And if they don't, how frustrated and angry will they be? And what will they do with their anger?

Giving up what we have and turning it over to them will not be an answer, because it won't go nearly far enough and will soon be gone again. We do have to use less, waste less, and pollute less. But the world's poor billions have a job to do, too. They must imagine a new world and work hard for change. They must control their population explosion and learn to practice tolerance and democracy.

And we must give them the tools—not just the fish but the fishing rod. We must find the best, most forward-looking people in those countries—every country has them—and strengthen their hands. Sure, they have different cultures, but the idea that some cultures or religions or people cannot be democratic or successful is just a kind of bigotry. It would condemn those people to never achieve their dreams.

Unfortunately, force has a role to play. It must be used very carefully, but sometimes it is the best choice. We have seen it work to positive effect in the Balkans and Afghanistan. We have seen it unite different nations in a common effort against truly bad men with savagely harmful ideas. We saw it half a century ago, too, as Germany and Japan rose from the rubble of their shattered dictatorships to become two of the strongest and freest democracies in the world.

But to rely only on force is to drift backward in history toward that bad old time when might made right. If, as I believe, our way of government is superior, then most people will not have to be forced to adopt it, especially in this age of instant worldwide communication. Why is it superior? Because it accepts the human differences others try to abolish. Because it respects the individual while enabling the community to act more or less rationally and fairly. Because it accepts differences of opinion and enlists the very energy of conflict to create new possibilities. With the right kinds of citizens, democracy works.

Of course, it goes wrong in many ways, and we constantly fall short of our own ideals. But we can get steadily closer, and the best hope for improving democracy is democracy itself. Indeed, bad as it is, it is not just the best form of government, it is the only form that expresses human nature. That is why, for the human species to last, democracy must spread throughout the world.

NOT JUST TECHNOLOGY BUT SPIRIT

However, with our freedom, we can destroy the world, and when we refuse to consider changes in our way of life—the over-consumption, the waste, the garbage, the arrogant displays of wealth—we place ourselves on a path toward destruction from which democracy alone cannot save us. To survive we must somehow combine freedom with self-restraint. This will be something new in human evolution, and it requires a spiritual change.

We evolved in face-to-face groups of close relatives, but now we must view the whole world as family. We battled nature to survive, but now to survive we must protect it. The stranger over the next dune or beyond our patch of forest was simply a threat to be fought, but now that ancient view is too simple and dangerous. For almost all our history, we solved problems day-to-day or at most year-to-year, but now we must think ahead centuries. For ten thousand years we tilled the family farm and cherished it, but now the family farm is the whole earth.

Technology is part of the answer. We need new ways to produce food and energy, create fresh water, defeat diseases old and new. We need the very best science just to figure out what we are doing wrong. But technology is also a threat in more ways than one, not just because of the physical damage it does, but because it can raise false hopes.

The science of genes is a wonderful thing, and someday it will lead to an understanding of health and illness, including mental illness. Perhaps we will even use genes to change human nature. But that possibility is very far off. We must solve our current problems with the genes we have now. A manned voyage to Mars would be a bold move, but if we become less determined to save Earth because we think we can get off it, it is incredibly dangerous. In the time frame that matters, we are stuck on this planet, and we had better make the most of it.

In DNA sequence, we are almost 99 percent chimpanzee. In 7 million years, we first walked upright, then expanded our brains and made tools to conquer nature. We came out of

Africa twice and filled the earth with our ambitious selves. And then one day, after countless adventures, after endless joys and sorrows and bravery and fear, we stood in a torch-lit cave in awe of the great paintings made by our kind, and we knew at last that we were really special. Not just in technology but in spirit.

Now we must draw on that quality, unique in the animal world, of spiritual consciousness and awe in the face of beauty. We must have the highest possible vision of our future on this small, beautiful planet, and meet all our challenges with that vision intact. Awareness, responsibility, resolve, courage, compassion, fairness—these have always been part of our nature even when they were swamped by fear and selfishness. Now, we have to save the world. The good news is, we can. Will we? Time is running out. In the words of one leading environmental expert, it's God's last offer.

If we do get past this bottleneck, we may emerge into a Golden Age where population declines of its own accord—as it has in many countries in the world once education, quality of life, and child survival became good enough. Once it becomes possible, people want to have fewer children and give them better lives.

After that trend spreads throughout the world, human numbers will be less and less of a threat in each generation. Your grandchildren will be looking at a future very different from yours, one where people can live well, protect the planet, and have enough resources to share instead of fighting over. They may look back on us and scratch their heads, wondering why we couldn't figure it out sooner. But you can be one of the ones who did!

---

* An earlier version of this essay appeared in Arthur B. Shostak, ed., *Viable Utopian Ideas: Shaping a Better World* (Armonk, N.Y.: M.E. Sharpe, 2003), 237–244. ("Our Future as a Species: A View from Biological Anthropology")

THE FUTURE OF US 159

FURTHER READING

Supporting evidence for this essay can be found in the author's book, *The Tangled Wing: Biological Constraints on the Human Spirit*, published in a revised edition in 2002. References are on the World Wide Web at www.henryholt.com/tangledwing. Those about human evolution are mainly the ones for Chapter 3, "The Crucible," while those about the environment and the future are mainly the ones for pages 463–481. *God's Last Offer: Negotiating for a Sustainable Future* is by Ed Ayres (New York: Four Walls Eight Windows, 1999).

WEB SITES

www.mnh.si.edu/anthro/humanorigins/
The Smithsonian Institution Human Origins Program. Up-to-date scientific information on human evolution.

www.prb.org/
The Population Reference Bureau home page. Current data on the status and growth of the human population.

www.worldwatch.org/
Probably the best, most up-to-date, and most readable accounts of the state of the environment.

www.un.org/esa/sustdev/natlinfo/indicators/isdms2001/table_4.htm
United Nations Division for Sustainable Development framework for social, environmental, and economic assessment of nations. Many links to individual topics.

www.grida.no/soe/index.htm
United Nations Environmental Programme—Global Resource Information Database [UNEP/GRID]. Everything you wanted to know about the state of the world's environment but haven't gotten around to asking.

www.cartercenter.org/peaceprograms/peacepgm.asp?submenu=peace-programs
The Carter Center in Atlanta actively works toward peace, human rights, and democracy around the world. This site describes how the center does it.

160 MOVING ALONG: FAR AHEAD

www.findarticles.com/p/articles/mi_m1252/is_6_136/ai_54422892
Democracies may go to war but not against other democracies.
This article tries to explain why.

Appendix

STUDENT FEEDBACK

Sixteen high school volunteers read more than sixty candidate essays, and offered feedback on many (though not on all) that influenced the final selection. Their (anonymous) views below are listed in the order in which they arrived back to me. They join me in hoping this material helps you take more from the essays.

A DIGITAL DAY—*JAN AMKREUTZ*

1) Although this essay does provide an interesting glimpse into the future, I don't feel that it really connects with any issues at hand. I think that this narrative would be better replaced with an essay like the one about the economical benefits of space exploration and colonization.

2) This essay is incredibly creative and original. I've read a fair share of "in the future" short stories, and this one gives the most in-depth explanation of how technology *will* really innovate a human's life.

However, I find this to be the main fault of the essay; it seemed never-ending for me. It simply didn't hold my attention. By the middle, I was already tired of hearing about all the new technology and gadgets.

These ideas are incredibly interesting and definitely relevant to people of this generation but would be better if they played into more of an actual short story, instead of one man's day.

If the main focus of the essay wasn't solely the new technology that lies ahead and instead had more of a "dual" plot to go along with the story of this man's life, it would be a better read. Maybe the man's life could be elaborated on.

3) Overall this essay is good; however, I do believe it needs some changes. It needs to be clearer about who is speaking. It is hard to follow because you don't know whether the protagonist is thinking to himself or talking out loud or whether a machine is talking back to him.

Also, a lot of information is presented in the essay. Too much of this technical information is boring, and it loses the reader. Another uninteresting thing is all the talk about insurance, taxes, the stock market, finances, and social security. It confuses readers, and most teens do not understand these issues yet or care about them. When the essay refers to real estate, kids cannot really follow because they don't understand it.

It would also be nice to know when this is taking place; kids want to know how close or how far away this is from the present.

There are some things that I do like about this essay. Some parts are humorous and very clever. It interests readers because it makes them wonder if the future will really be this way.

Overall I like the essay. It is amusing to see how different the world could be in the future. The only major criticism I have is that teens may not be too interested in what the future will be like for adults. They want to see what kind of change there would be in their own lives right now. Teens are not too interested in their lives fifteen years from now.

4) I think you should definitely accept it. It really kept my attention, and the imagery helped me to picture it. The details helped me to really create the scene in my head.

I liked that it described the technology, but it didn't get too technical. It remained a work of literature rather than turning into a scientific report. I thought it was really interesting and really imaginative.

Everyone thinks about what it will be like in the future, but the author really thought it through and made it come to life. He expresses his ideas clearly and effectively.

5) This essay should be included in the book! It has a different insight into life in the future and includes many aspects that could soon exist.

6) Although the author of the essay "A Digital Day" obviously has a great deal of creativity and potential, I found this particular essay underwhelming. It was too similar to all the other science fiction stories out there. It was overdone and

emphasized unnecessary details to pound into the reader's head that it really was a sci-fi story.

7) This essay was interesting but very long and a little boring at times. Shorten it, especially when it goes into the specific details about old companies and real estate. Emphasize the things that are more interesting like the shoes, cars, and digital twins. Who is viewing the information that the shoes put out and how? What do these people do for entertainment? Make it flow better. Be sure to keep the ending with the questions.

8) With some editing, I would accept the "Digital Day" essay. It is interesting (of course, it would be better to read *Neuromancer* by William Gibson). It is easy to read, though not spectacularly well written.

FUTURE HEROES 2035: THE BIG PICTURE—*JOHN SMART*

1) I do not really see the difference between this essay and the "High Schoolers Tomorrow" essay. They are basically the same. I am not a fan of either of them. They are long, and they often lose the reader.

2) I liked the flow of the sentences and the language. It was a little long. Maybe the author could cut out the part about outer space. I liked the part about your electronic self and the evolution/development. The concluding paragraphs are a little wordy and the author could end it a little faster.

3) I like this essay, but it is not one of my favorites. I agree that it is too long, though. About two-thirds of the way through, I lost interest. It was hard to keep my focus, simply because it was hard to take in all the new ideas that occurred in this future. I would cut down on the number of new ideas. These ideas are very *cool* and unique—which is what I find particularly interesting about this essay; but toward the end it was a bit overwhelming. I would cut down on the emphasis of the biological future. Also, I think that I would cut the paragraph that starts: "The theories in Evo-Devo ..." If it is shortened, I think it would appeal to many teens.

4) I "voted" not to include this essay in the beginning, so I won't be of much help now, but it is a fair amount shorter.

5) Personally, I think this is fantastic. The concepts mentioned in the essay border on science fiction but are also very foreseeable. Some parts of the essay would go over some people's heads, but the majority of it is written in language that is easy to read and comprehend. The first-person, realistic tone makes it very believable. Also, the sources at the end are very nice. I recognize the "ascent of man" because it was the focal point of our English class last year, and I have also read parts of *Cosmos* and *Connections*.

6) I do not really like this essay, though I did like the beginning. It caught my attention with the whole slang vocabulary, but then it quickly lost me. There is no real action in the essay, and it is just someone talking the whole time. I do not think teens will find this essay too interesting.

7) I enjoyed this essay. It is a *complete* futuristic world, which always leads to the most intriguing storylines. Although some of the concepts and ideas are overused and even dated by today's standards, I think it is worthwhile to have at least one essay of this kind in your published volumes. In a way, no science fiction is complete without a little piece like this.

8) I liked the predictions and how it went back to the current century, but it was way too long. I lost interest about halfway through. I liked the beginning about what culture may be like, but I really disliked how he tried to use cool words throughout the essay. It seemed like the author was trying too hard because the placement of the words was kind of random.

9) I never really liked this essay. It is really boring having someone "talk" to you the whole time. It is not very interesting. I think the other essays I read should be in a book before this one.

LETTERS TO UNBORN DAUGHTERS: EXPLORING THE IMPLICATIONS OF GENETIC ENGINEERING—*SARAH STEPHEN*

1) Although this essay provides an interesting glimpse into

our future, I don't see how it could keep the attention of most high school students. Perhaps the shorter version cited at the bottom of the essay would be a better choice.

2) I would accept this one. I personally found it very interesting and very unique in style. Having it in a letter format from mother to unborn daughter adds a more intriguing spin to a very thought-provoking aspect of the future.

3) Genetic engineering has always fascinated me. However, I don't approve of selecting factors to make a child "perfect." It would be great to get rid of genes that caused terminal illnesses like cancer or heart disease. It would not be right to select IQ, eye color, height, etc. It messes up the natural balance of nature and human companionship. It would also eliminate the class of people who are dedicated to jobs that require normal intelligence, for example, secretaries, trash men, janitors, and factory workers.

Genetically enhanced people would eliminate the need for hard work because no one would have to "try" at anything. There would be no surprises left. Everyone would be perfect, and living life would be almost a waste because there would be nothing to work for if everything was given to you on a silver platter.

I think this essay is very important for people to read. If and when we get to the point these mothers were at, the unnatural tendency to genetically enhance babies would be the end of life as we know it. The human race would be separated, not phenotypically, but genotypically.

"I remember wishing my parents had selected something to be "turned on" when I was created. They didn't, so I had to work hard to be good at soccer. But I think I was a better player for it. I practiced harder on my technical skills than some of the other girls, but I still made the varsity team, so really, I wasn't too far behind. In the end, I'm glad that your grandparents decided not to provide me with any extras. I know that I was successful because of my hard work, not because science made it easy for me."

This daughter, who was not genetically enhanced, completely understood the value of hard work and her own achievements; she is one of the deserving. The rest never had to experience this; they are undeserving. Eventually, if people are all genetically enhanced, then there will be no "try," just simply "do."

"She never considered that I didn't want to score high on the tests and flunked them on purpose. And imagine her frustration when I grew to be three inches shorter than she expected. She was furious. I'm glad that everything didn't work as she planned. I hated her expectations and her anger that I wasn't as perfect as she had asked for me to be. I was alive and healthy; isn't that perfect enough?"

There will also be problems with high expectations. People who are naturally gifted, whether in athletics, music, or academics, already experience an alienation from everyone else. There are many expectations for such people, and it frustrates them. They are unable to do what they want just because they want to. They have to do what is expected, so as to not disappoint society or their family. That is not what it should be about; it should be about doing what you want to do, because you want to do it.

Already we have a problem with unusually gifted people. What happens when almost everyone is that way? There will be no differences, and competition will lessen. It will lessen, because people won't have to work for anything, and they will not need to fear failure.

This essay should be published, because someone, somewhere, might see the logical dangers that genetically enhanced, "perfect" humans could do to the condition of the human race.

4) I think the topic about genetic engineering is good to put in the book. With so many advancements in technology, this type of science is becoming more and more realistic and right around the corner.

However, I don't like the style. It seems boring being written in separate letters. I think it should be more like the essay on the pill (the one you take to get smarter). It should show the positives and negatives about genetic engineering and let the readers decide what side they want to choose.

5) I really like this essay. I like how it sets the scene and

continues on with the same family showing the consequences of each person's actions and how they affect not only themselves and their own children but also the world around them. I enjoyed reading it, other students will too.

6) I really liked this essay; it is one of my favorites. It held my attention the entire time. The essay addresses a facet of this issue that I've never considered even looking at, the competition between "bests" and natural born people. It almost instilled fear in me for the future and the huge role that genetic engineering could play. I liked this essay especially because of the format. The letters were an interesting way to convey how genetic engineering could destroy social relationships and instill competition into young children and their parents. I would include it.

IT'S TWELVE O'CLOCK, AND I KNOW EXACTLY WHERE MY YOUNGSTERS ARE—*JOHN CASHMAN*

1) I would definitely include this essay. The format is well put together because of the combination of the story and the facts. This essay held my interest all the way through, especially because of its short length.

2) I am not too sure if kids will like this essay. I think a parent would like this essay more to see what they would do to keep track of their children in the future. This issue has not even really passed kids' minds yet. Teens hate when their parents spy on their every move, and they might hate this essay because of it. I just do not think kids my age are concerned about their parenting in the future.

BIOLOGY IN YOUR FUTURE: PREPARE TO BECOME SOMEBODY NEW—*MELVIN KONNER*

1) I like this essay. It shows teens what to look forward to in the future and what different types of consequences the future could bring. Also, I like how the author does not use far-fetched topics. The issues he writes about are either here or right around the corner. Teens will find a lot to interest them in this essay.

2) I like that the essay, on the surface, is about biological advances. But as you read it, it seems to touch on a lot of other areas as well, without over-explaining them, or losing sight of its original goal.

3) You should include this essay. I liked that the predictions were very plausible based on the current technology and that it provided a little warning at the end. I liked the way it was written, and it held my attention. This was one of my favorites.

4) I liked this essay because it gave insight into a wide range of aspects of life (from genetically modified foods to STDs). The only issue is that it is a tad long. I don't have an idea of what should be cut, but it should be shortened a bit.

NANOTECHNOLOGY: BIG REVOLUTION WITH SMALL THINGS—
JIM PINTO

1) Nanotechnology seems incredibly cool. It is undoubtedly part of my future. The article was fun to read. I have a particular interest in microbiology and biochemistry, so some of the information applied to my interests. My only question is that if we make computers as small as a sugar cube, won't they be lost? I think it is great for disease prevention and whatnot, but people are getting too hyped up on minimizing everything. The article is definitely a good read and deserves a place in the series.

... AND THE BUBBLEGUM POPS: NANOTECH VS. CAPITALISM—
GLENN HOUGH

1) This essay would be good in the beginning of the book. I really liked the quotes, and how the author gave the definition of nanotechnology and said how it could be applied. It was interesting but a little scattered. The definition should go at the beginning and you should include it.

2) Wow, this essay just blew my mind. The essay is excellent; it draws you in and holds you. It doesn't get too technical, but it opens up the imagination and the mind. Include it by all means. I really enjoyed it.

3) I would include this essay, but it is not my top choice. I would include it because the writing speaks to the reader—which I think is something every teen wants, and the topic is interesting. The quotes at the beginning bring the reader in because they're almost absurd, and I wanted an explanation of them. The essay is to the point and doesn't become side-tracked.

4) I like this essay, but I like the first nanotechnology essay you sent me better. This essay needs more examples of what nanotechnology will do for us in the future; what kinds of things teens will use in the future as a result of nanotechnology (cell phones, Internet, television, etc.)

SPACE: TEENAGERS AND THE FAR OUT—*JEFF KRUKIN*

1) Space and space exploration are topics I have always had an interest in. I think that especially with the Mars rover landings we are going to see a rekindled interest in space; in fact, the author quotes President Bush's "A Renewed Spirit of Discovery." This essay will appeal to a vast majority of students because it is a topic that is growing in relevance to their future.

2) I really like this essay. Space has always interested me and gives off a sense of wonder. Teens will enjoy this because it is right around the corner. They also have a reason to be interested in it because of the announcement of President Bush's new program. It shows teens what to look forward to, what to expect, and what to do to make the future of humans in space possible.

3) This one dragged on. I think that space exploration should not be a government franchise. It seemed overdeveloped.

4) I really liked this essay. It caught and kept my attention. Space has always been a faraway concept for me. I've never really just thought about space and how I could contribute. I like the way this essay lays it all out and gives plenty of examples of the endless possibilities for contributing to space exploration. It was very well written, and I think it will keep the students' attention.

ADDENDA TO SPACE: TEENAGERS AND THE FAR OUT

1) I think the original essay is fine by itself. The author's points were expressed, and, no matter whether counterarguments are printed or not, this kind of debate will always exist. I wouldn't include it but rather let people argue it out for themselves.

2) I really like the comments to the essay. I think it is a great idea. Teens will like seeing the different viewpoints and choosing their stance.

3) While the opinions expressed are very interesting, a debate between two authors is what the book is really supposed to include. Since it is supposed to include the opinions of students, it doesn't make sense to include the Addenda after the essay. The students who read the series might not be able to produce the exact questions proposed in the Addenda, but they will be the type who will come close. Their reasoning will be along the same lines. The students reading the essay who are passionate about the topic will research it—or already have—and will be able to produce counterarguments. Some may appreciate the opportunity to read the Addenda, but it does not fully follow the context of the books.

DEBATE: SPACE AND OUR NEAR FUTURE—*JEFF KRUKIN*

1) I like this essay. There were many improvements to it, and it reads well. This is an important issue in our future, with the date of landing a man on Mars projected around 2020.

2) I liked the original, but the revised version is even better. Hopefully others will find the new one worthwhile as well because it is an interesting piece that should be included.

3) This essay is more valid and less biased now that more information was provided on the other side of the debate.

4) If this were a debate, then it would be more equal on both sides. It would not only have one-liners from the anti-space end. This is a hot issue that has two well-developed arguments, not just one. If this is published, the other side should have more backing, or this essay would only serve to brainwash, not allowing people to develop their own opinions on the topic.

5) While I like it and think that it should be printed in some large magazine like *Time*, I don't know if it would fit that well in the book. First, there are several essays on the same subject already in the mix. Second, it fits, as did the postwar Iraq article, more in a book about the near future than in one on the progress of our race and world. While the essay supports a valid argument for space exploration, and the feasibility of Bush's plans, there are more appropriate places it could be published.

6) It was an in-depth look at one side of an argument. What about the other side? I think that it is too early to perform too much space exploration. We don't know enough about the sciences.

7) I think this essay is too much about President Bush's program for space. It seems too political to me. I do not think teens will be very interested in this.

8) This is a real big negative with me. I do agree with space exploration, and I am a firm advocate of Mars exploration and eventual colonization. I even consider the extension of the human race to Mars to not only be another great step forward but also to be necessary for our overall survival.

The time will come—so far off it is not even really worth mentioning—when the Sun will cease to exist in a stable form for human life. At this time, we will either perish or live safely in another solar system. As crazy as that may sound, it is only probable if we put more effort into it.

In the short term, we live in an irrational and unstable world. With so many nuclear weapons and disagreeing factions, Earth's very integrity is at stake, as one great war could finish us all (MAD). By colonizing Mars, our chances of survival double. Is it worth it, yes!

As for this essay, I did not find it either worthwhile or logical. To proclaim the exploration of space as noble is one thing but to do so by manipulating certain political truths is another. Fact is one thing; speculation is quite different. This paper pursues one track of thinking and because of that tunnel vision, it limits its prospective audience and causes the reaction I am now voicing. I would reject this one.

9) I really liked this essay; it was informative and interesting. The author defended his points very well. He cited important figures. It might be interesting to get the other point of view, though—maybe if there was another essay out there that talked more about the reasons we shouldn't use part of the budget for space exploration.

10) I'm not quite sure why, but I didn't really like this essay. Maybe it was because I disagreed with what he said and the way he said it. But that's just my own personal opinion and not necessarily the opinion of others. I like the way he put out the statement and then ripped it apart, but he was a little too—not necessarily sarcastic—but a little too strong with his humor. Anyway, I think that some students might like this essay because they might agree with him and like his style, but I didn't.

WHY SHOULD WE SEND HUMANS TO MARS?—
THOMAS GANGALE

1) I like this essay, too, but the one you sent me earlier about the program Bush announced was better. I like it more because it tells kids why they should be interested in space and what it could bring to our own future. Perhaps you may be able to combine the information from both essays.

2) The essay was good, especially the conclusion. It is an intriguing topic, and the essay kept my attention the whole way through. I enjoyed how it went back into history and came up with big names that we will recognize and relate to. Students would like to read this essay and would get some good insight from it.

3) I would approve this one. I really enjoyed it when I got to this line: "But the images of the Earth that we brought back from the Moon are timeless and universal, because they are the first images of all of us." That is a sentence that will stick with people, not to mention the message or subtle call to arms of this entire essay.

Mars is an opportunity that is lying dormant, waiting to be seized. It is time we as a race took advantage of our resources and crossed into the next frontier. Essays like this may help turn the tide into a future of progress and, in this case, colonization.

4) This issue is very pertinent to the future of today's students. However, including this one may be going a little overboard with the space topic, so I would include excerpts.

5) I read this essay once and thought it had a very good point. The next day I was quoting it to people because I liked its point so much. Not only does it cover the scientific topics, it also covers the socio-dynamics of the subject, which are perhaps more meaningful and important overall—a fascinating essay that I would include.

A MARS COLONY: C096—*JOHN A. BLACKWELL, PHIL GYFORD, GLENN HOUGH, ALEXANDRA MONTGOMERY, AND DANA WILKERSON-WYCHE*

1) I do not like the essay you sent to me about the blueprint for the Mars colony. It seemed to me like I was reading a constitution. Many teens do not even understand how our government works; let alone one that is entirely made up. They are not interested in politics in the future. They want to see technological advancements.

2) I liked the essay, but it was a little lengthy. Maybe it would be better if it was cut down a little and used a little less detail. For example, I liked the section about the passage of the bill—about how it was a multiple-choice process. Then it got confusing.

It was an interesting essay because it shows a combination of all the forms of government we have at present and how maybe in the future they will decide that no one form of government is perfect and instead combine elements from each.

There were just too many parts. To include the day in the life, the Code of Ethics, the Bill of Rights, and the design principle is just too much. It didn't really add anything. Maybe a clearer introduction could be written, too. I had to read it a few times to figure out what was going on.

3) The idea of this essay is great, being a synthesis of sci-fi and politics. The vision behind this is visible, and while futuristic,

still appears tangible. The appendix adds to this greatly, enhancing the idea of the government.

I don't think the "Day" epilogue adds anything particularly, however, and I would drop this from the essay. The first-person example is nice, but the information in the main essay and appendices provides a clear enough viewpoint of the system in my opinion.

ABSTRACTS FROM *FUTURE SURVEY*

Michael Marien, editor

The following abstracts were prepared by Michael Marien, founder and editor of Future Survey, *a nonpartisan monthly newsletter published by the World Future Society since 1979. FS provides fifty abstracts every month of recent books, reports, and important articles on global and domestic issues. FS carries items on trends, forecasts, and policy proposals on topics such as world futures, the global economy, the Middle East and other regions and nations, security, energy, the environment, governance, education, health, crime, communications, new technologies and their impacts, and methods for shaping a better future.*

TECHNOLOGY/A.I. (September 2003)

When Machines Outsmart Humans. Nick Bostrom (Department of Philosophy, Yale University and Oxford University). *Futures,* 35:7, September 2003, pp. 759–764.

Any scenario of what the world will be like in 2050 that simply postulates the absence of human-level artificial intelligence is making a big assumption that could well turn out to be false. It is thus important to consider the alternative possibility: that intelligent machines will be built within fifty years. Indeed, considering Moore's Law ("not a law at all," but merely an observed regularity), *"human-level computing power has not been reached yet, but almost certainly will be attained well before 2050."* We know that the software problem can be solved in principle. We are beginning to understand early sensory processing. There are reasonably good computational models of primary visual cortex, and we are working our way up to the higher stages of visual cognition.

We should thus consider the ramifications of human-level artificial intelligence in the mid-twenty-first century: 1) artificial minds can be easily copied (the marginal cost of creating an additional artificial intelligence after you have built the first one is close to zero; *"artificial minds could therefore quickly come to exist in great numbers"*); 2) human-level AI leads quickly to greater-than-human-level AI (this essential point makes AI a truly revolutionary prospect; if Moore's Law continues to hold in this era, the speed of AI will double at least every two years); 3) technological progress in other fields will be accelerated by the arrival of AI (a true general-purpose technology); some have speculated that this positive feedback loop will lead to a "singularity" where technological progress becomes so rapid that genuine super-intelligence is attained (however, there may be diminishing returns in AI research when some point is reached); 4) unlike other technologies, AIs are not merely tools—they are potentially independent agents.

This lead essay in a symposium entitled "Machines Are Us?" is followed by five responses: 1) J.R. Ravetz (Research Methods Consultancy, London) warns of the potentially evil effects of real AI, and encourages thinking about the sorts of things we want to avoid; 2) Steve Fuller (Warwick University) doubts that improved machines will be accorded the legal and moral status of human beings and attacks various failings of "futurologists"; 3) Graham T.T. Molitor (Public Policy Forecasting) questions Bostrom's timing, embraces Ray Kurzweil's assertions that AI will be accomplished before 2019, and suggests that human brains may also develop (in that humans use less than 10 percent of brain capacity); 4) Anne Jenkins (University of Durham) critiques Bostrom's narrow view of intelligence that overlooks the role of emotion, experience, society, and a philosophical framework, and doubts that autonomous AI can operate in the real world; 5) Rakesh Kapoor (Alternative Futures, New Delhi) warns that science applied carelessly can lead to unmitigated human disaster, and asks whether developing beyond-human-level AI is desirable.

TECHNOLOGY/GENETICS (September 2003)

The Double-Edged Helix: Social Implications of Genetics in a Diverse Society. Edited by Joseph S. Alper (professor of chemistry, University of Massachusetts-Boston) and five others (all

from the Genetic Screening Study Group). Baltimore, Md.:
The Johns Hopkins University Press, October 2002, 293 p.

A group of Boston academics, first called the Sociobiology
Study Group and renamed the Genetic Screening Study
Group, began meeting in the 1970s to discuss interpretations
of human genetics. This book expands explorations of the
impact of genetic research on contemporary social life by high-
lighting how nondominant groups understand and are affected
by genetic technology.

Essay topics: 1) genetic complexity in human disease and
behavior ("*the simple notion that there is a single gene or several
genes for a particular trait like intelligence or homosexuality is clearly
erroneous*"; in practical terms, "we cannot expect simple solutions to genetic problems"); 2) geneticists in society (no one
can ascribe any particular stance on eugenic practices to those
doing genetic research; nevertheless, "*the view of geneticists
most often heard by the public tends to overestimate the role of genes
in human affairs*"); 3) advocacy groups and the new genetics
(many support groups play a critical role in advancing gene
discovery with the hope that better knowledge will bring
improved treatment, but "*genetic research can be a double-edged
sword for people living with genetic conditions*"); 4) gender and
genetics (women are often strangely neglected and invisible in
the debates about the potential uses and abuses of new genetic
technologies); 5) pre-natal diagnosis and selective abortion
(with about 50 million Americans having disabling traits, it is
important to pursue a policy that it is as acceptable to live
with a disability as it is to live without one, and that society
will support everyone with the inevitable variety of traits, and
thus diminish the desire for selective abortion); 6) African-
American perspectives on genetic testing (generally more
mistrust of the motives of genetic medicine, with the link
between risk and stigma becoming highly intensified in contexts of racism and poverty); 7) the Human Genome Diversity
Project (an undertaking of great scientific interest, "it is
unlikely that this knowledge will be of material benefit to the

indigenous peoples who will be studied" and geneticists must be wary that their knowledge may be used by racists); 8) the search for a genetic cause of homosexuality (researchers have an ethical obligation to consider the potentially negative impact of their work); 9) commercialization of genetic technologies (hopes for biotech should be tempered by concerns about discrimination, psychosocial harms, and erosion of public trust of science; 10) current developments in genetic discrimination (insurance companies, clinical professionals, adoption, armed services, employers, and blood banks; discrimination does exist, and the public believes it to be a threat to their well-being).

[NOTE: Solid critiques to keep the one-eyed enthusiasts of genetic research on a needed leash.]

TECHNOLOGY/NANOTECH (September 2003)

The Next Big Thing Is Really Small: How Nanotechnology Will Change the Future of Your Business. Jack Uldrich (former deputy director, Minnesota Office of Strategic and Long-Range Planning; www.nanoveritas.com) with Deb Newberry (Burnsville, Minn.). New York: Crown Business, March 2003, 207 p.

Materials 100 times stronger than steel and one-sixth the weight are "not the predictions of some wild-eyed futurist," but the conclusions of serious scientists describing the potential of nanotechnology: the willful manipulation of matter at the atomic level. Citing an April 1, 2002 article in *Business Week,* *"before 2010, the market for nanotechnology products and services is expected to reach $1 trillion in the U.S. economy and will require anywhere from 800,000 to 2 million new jobs."* Every major government in the world is now investing in nanotech, and big corporations are outspending government by a two-to-one ratio. A dozen universities have established multi-million-dollar nanotechnology centers since 2001, and dozens more are expected to do so in the years ahead.

Expectations are grouped in four near-term time periods: *2004 & 2005: Faster, Smaller, Cheaper, Better:* a variety of diseases better understood (Alzheimer's, cystic fibrosis), better disease detection for many types of cancer, nanosensors to detect the presence of anthrax and a variety of illnesses (ten times faster than anything today, and 100,000 times more accurate, reducing the need to send material to a lab and the number of lab technicians needed), nanosensors to detect poison gases like sarin, "smart bombs" to treat cancer (a single atom of radioactive material in a nanoscale cage made of carbon and nitrogen with a protein that attaches only to cancer cells), nanocrystals to mix with the body's own cells to help regrow bones and treat osteoporosis, a nanoscale device to encapsulate an asthma drug and get it to the lungs more effectively, a nanotech boost to effective fuel cells, thinner and more powerful batteries, a market for nanomaterials growing at 300 percent a year, reduced cost of carbon nanotubes from $500 a gram to less than $6 a gram, nanofilters to make impurity-free drugs and purify water;

2006–2008: The Avalanche Begins: Many industries can expect to see their economic landscape transformed (semi-conductors, publishing, advertising, food, clothing), plastic semiconductors to enable an "electronic paper" modified to look and feel like a newspaper (much more convenient than e-books), use of electronic billboards with changing messages, more accurate longer-term weather forecasting, stronger and lighter materials making blimp cargo lifters a viable option for some transport needs, nanoparticle taste enhancers to add to low calorie foods (likely to shake up the $42 billion diet-supplement market), nanoparticles neutralizing harmful chemical agents for environmental cleanup;

2009–2013: Taking Control: a nanotech remedy for drug abuse and addiction, nanoscale dendrimers (tree-branch-shaped devices to carry different chemical tools to locate and kill a cancer cell), nanosensors painlessly inserted into one's arm to test blood for illnesses, deployment of nanosensors at meat-packing plants and at border crossings to detect food-borne illnesses, fuel cells enabling homes and businesses to become net energy producers, nanotech lighting advances to reduce world energy consumption by 10 percent;

2013 & Beyond: new materials to enhance space exploration, nanotech-enabled solar cells brushed in layers on almost any product that requires power, growing new body tissues to address the severe shortage of organs for transplant (nanotech and other medical advances will increase the number of people over 100 years old, and "push the upper limit to 120 years and beyond"), nanoelectronics-based computers approaching the capacity of the human brain, nanofiltration systems to desalinize water (thus avoiding many global conflicts).

[NOTE: Popularized survey by a somewhat "wild-eyed" consultant but worth considering. If only half of this comes true even in twice the time foreseen, it will still be huge.]

TECHNOLOGY/SPACE (September 2003)

Space: The Free-Market Frontier. Edited by Edward L. Hudgins (former director of regulatory studies, Cato Institute). Washington, D.C.: Cato Institute, December 2002, 259 p.

What has happened in the last three decades to delay humankind's full exploitation of space, and what can be done to speed things up? To move from the current situation of limited access to space and to truly make space a place for humans to work, live, and play, we should consider how we arrived at the current situation, the promise of a commercial market future, and what policy changes might make space the next commercial market frontier.

Many of the problems of the space sector can be traced to the different paths taken by civil aviation and space flight. When the Wright Brothers made their first flight in 1903, they acted as private individuals, pursuing their own vision with their own money. The saga of space flight started much as civil aviation did, when Robert Goddard launched the first liquid-fueled rocket in 1926, funded principally from a private foundation. The U.S. space program was driven by the Cold War, and NASA has pretty much controlled space.

A market-based space policy would benefit those who seek

lower costs for access to space and more space infrastructure and services. Private entrepreneurs are providing many services or trying to develop marketable services and missions. Examples include the Cosmos I space sail pushed by the solar wind, the RadioShack rover on the Moon, services provided by SpaceDev and SpaceHab, the design for space stations based on external fuel tanks (worked out by Space Island Group), private launch services (offered by Lockheed Martin, Kelley Aerospace, and Kistler Aerospace), the plan of Bigelow Aerospace to build and orbit a private space station at a much lower cost than the International Space Station, the possibility of placing large solar energy collectors in orbit, and Robert Zubrin's low-cost Mars mission proposal [*FS* 22:3/108].

Essays discuss reducing the federal government's presence in space policy, prizes to encourage private space flight, NASA in the twenty-first century with entrepreneurs as allies, barriers to space enterprise that create uncertainty and added risk, a legal regime for private activities in outer space, a proposed multilateral treaty on jurisdiction and real property rights in outer space, the structure of the space launch market, spreading the benefits of space travel to society, space tourism and space hotels as a catalyst for a new space development strategy (Buzz Aldrin and Ron Jones write of recent studies, including NASA's own research, suggesting that tens of millions of U.S. citizens want to travel into space), and a proposed unified theory of space property rights.

[NOTE: Many imaginative ideas that appear to deserve consideration.]

TECHNOLOGY/SPACE (September 2003)
Space Policy in the 21st Century. Edited by W. Henry Lambright (professor of political science, Maxwell School, Syracuse University). Baltimore, Md.: The Johns Hopkins University Press, January 2003, 283 p.

The launch of Sputnik in 1957 makes the Space Age more than

forty years old, but the most challenging years of space technology and policy lie ahead. Themes of the 10 chapters: 1) the challenge of space access (the first decade of the twenty-first century has enormous potential for developing new launch vehicles that are reliable and affordable, e.g. the Kistler K-1 reusable launcher); 2) the future of space commerce (the most important risk category and the most difficult to manage is political risk from inconsistent and counterproductive government policies; policy precepts are discussed, and three general futures are envisioned: the sky filled with commercial satellites and new markets in space tourism, space commerce mature but only a complement to terrestrial activities, and space commerce crippled by regulatory burdens); 3) the politics of earth monitoring from space (the number of uses for remote-sensing technologies has mushroomed, and trends will likely continue in the direction of a widening and deepening of remote-sensing data, presenting "a radically different environment for global governance"); 4) the Space Station era and international cooperation in the next decade (the optimistic scenario is "everything goes according to plan" and the central U.S.-Russia partnership remains stable; the pessimistic scenario is "things go wrong," especially an accident involving loss of life); 5) asteroidal threats and promises (four scenarios are described: scientific investigation of asteroid bodies, focus on avoiding collision, use of asteroid resources for space construction and development, military use of asteroids); 6) the quest for Mars (the von Braun vision [see above], new unmanned expeditions, current advocacy by Robert Zubrin, and three routes to Mars); 7) the search for extraterrestrial life (Europa and Mars as the most likely locations); 8) creating a new heritage in space (by John M. Logsdon, director of the George Washington University Space Policy Institute, who notes that political support for a substantial increase in NASA's budget is limited, but there is support for a stable budget); 9) the wondrous vision of those who imagined space policies some fifty years ago and how they have changed (*"these prophets wholly failed to anticipate the power of remote sensing, one of the great surprises of the Space*

Age"); 10) adapting NASA for the twenty-first century (the global level will become increasingly salient in agency decision-making; rivalry among space powers will continue, but within alliance projects).

[NOTE: A fine overview of problems and possibilities.]

WORLD/HUMAN RIGHTS (January 2004)

The Globalization of Human Rights. Edited by Jean-Marc Coicaud (Peace and Governance Program, United Nations University), Michael W. Doyle (adviser to the UN secretary-general), and Anne-Marie Gardner (Princeton University). Tokyo & New York: United Nations University Press, January 2003, 208 p.

The 1948 Universal Declaration of Human Rights, outlining "a common standard of achievement," has become the cornerstone of a burgeoning international human rights regime. Three trends highlight the increased prominence of human rights in international relations: 1) the proliferation in the number and scope of human rights instruments; 2) the regime's increased attention to implementation; 3) growing acceptance of the view that human rights places limits on state sovereignty (i.e., a state's legitimacy is tied to proper treatment of its citizens, and an offending state can no longer hide behind a mantle of sovereignty alone).

Essays focus on international ethics and human rights, relationships between civil/political rights and social/economic rights, incorporation of civic and social rights in domestic law, differences of human rights practices of the North and South, problems in implementing human rights in the South (bad governance, lack of law and order, poverty, the debt burden, lack of human rights education, government control of media, etc.), human rights and Asian values, the politics of human rights, transnational duties toward human rights and global accountability, and the continuing "uphill battle" for human rights.

To address demands of international justice and human

rights, at least three challenges must be met: 1) embracing and adjusting international diversity without smothering it (the question of how to implement a multilateral culture without having it become a tool of Western extension and colonization); 2) addressing the weak sense of international community (through stronger mechanisms of global identification and participation); 3) handling the paradox of contemporary democratic culture (the increased sense of responsibility at the international level and the simultaneous proliferation of a democratic culture of individual entitlement at the national level that is apt to be allergic to global solidarity).

WORLD/GOVERNANCE (January 2004)

An Essential Institution for the 21st Century: A United Nations Constabulary to Enforce the Law on Genocide and Crimes against Humanity. Saul Mendlovitz (professor of international law, Rutgers University; World Order Models Project) and John Fousek, *Annals of Earth*, 21:3, 2003, 14–16 (from Ocean Arks International, Burlington Vt.; www.oceanarks.org).

The signing by 139 states and ratification by 29 states of the statute to establish a permanent International Criminal Court with jurisdiction over the most heinous of violent international crimes (genocide, other crimes against humanity, war crimes) marks a moment of great promise, despite the statute's imperfections and the unlikelihood of U.S. ratification in the foreseeable future.

Human rights organizations and other citizen groups played an essential role in this treaty. A similar worldwide citizens' campaign to create a standing UN. Constabulary could have similar results in the decade ahead.

"With these two new global institutions in place, the prospect for enforcing international criminal law and deterring potential criminals in the future would be immeasurably improved ...(and) go a long way toward breaking the cycle of impunity that has fueled the continuation of genocidal violence in the half century since the Genocide Convention was signed."

The United Nations Constabulary would be a new kind of force, with assertive police powers. It would be a permanent, transnational institution, with members individually recruited as international civil servants and employed directly by the UN, rather than by their national military authorities. It would differ from UN peacekeeping forces (which have always been formed on an ad hoc basis from national military contingents) and from proposed rapid reaction brigades (similarly composed of state contingents).

The Constabulary would be dedicated exclusively to dealing with genocide and other crimes against humanity. It would be a police arm of the evolving regime of international criminal law, entirely distinct from any UN forces used to deal with invasions of one state by another, or to intervene in civil wars where genocide was not a major component.

An early warning system—an International Crime Watch Advisory Board—needs to be developed to provide guidelines for when the police force would be brought onto the scene. Bosnia and Rwanda provide perhaps the most notorious cases where early warning signs went essentially unheeded.

A standing Constabulary of 10,000 to 15,000 should be housed in perhaps three or more strategically located base camps. It should be outfitted and trained in the manner of a highly professional national guard.

[NOTE: A specific proposal that could prevent much violence.]

WORLD/GOVERNANCE (January 2004)

The Architecture of Global Governance: An Introduction to the Study of International Organizations. James P. Muldoon Jr. (senior fellow, Center for Global Change and Governance, Rutgers University-Newark). Boulder Colo.: Westview Press, January 2004, 322 p.

The world of 2002 is a sharp contrast to the kind of world many analysts and political leaders had thought would emerge after the Cold War. Initially, there was a naïve belief that the demise of the Soviet Union and Communism has ushered in a "New World Order" of peace and prosperity based on core Western values. But, over the 1990s, the commitment and unity fractured under the pressure of a string of events.

The end of the Cold War and radical shifts of power away from states to non-state actors were the impetus behind the surge of interest in the concept of global governance. The state-centric international order was coming undone, but there is growing evidence of a global system emerging out of this chaos.

A few fundamental characteristics of global governance have emerged: multi-polity of power and decentralization of authority, emergence of new forms of "governance without government" (see Rosenau and Czempiel, 1992), and structures that promote stability in the global system.

Topics discussed: the classical schools of political thought (from ancient traditions through the sixteenth century), peace plans and reform ideas of the seventeenth century to the nineteenth century, the idealist/realist debate in modern international theory, the evolution of international organizations, the three institutional pillars of global governance (the political domain of states, the economic domain of multinational corporations and industry associations, and the socio-cultural domain of nongovernment organizations and professional associations), and international organizations and the management of global change.

The search for new mechanisms of governance is characterized as "a duel between reform and revolution." The reformist position, dominated by the United States, considers the current international architecture basically adequate and emphasizes slow incremental change. The revolutionist position considers the architecture antiquated and emphasizes radical change to correct participatory gaps. "The fate of existing international organizations will be determined, in large part, by the side that ultimately prevails."

The capacity of international organizations to meet the governance needs across the three domains of governance has to be strengthening. "Arguably, contemporary terrorism and the other anti-systemic forces created by globalization are sympto-matic of the institutional deficiencies within all three gover-nance domains." They are not likely to yield to traditional power structures of twentieth-century international relations, "no matter how much firepower the U.S. has and can project."

[NOTE: Very dry and ponderous but useful background nonetheless.]

WORLD/2050 VISION (January 2004)

Matters of Consequence: Creating a Meaningful Life and a World That Works. Copthorne Macdonald (Charlottetown, Prince Edward Island, Canada; www.cop.com). Foreword by Paul H. Ray (coauthor, *The Cultural Creatives*). Charlottetown, PEI: Big Ideas Press, February 2004, 374 p.

Through the 1960s and 1970s, it became increasingly clear that "progress" was not a flawless boon for humanity or for other species. By the 1990s, global life-support systems were experi-encing many problems. "Awareness of these realities has now become widespread, and this has led many people to experience ethical discomfort and consequent calls for action."

We must transform some of our present modes of personal, social, and economic functioning into modes that are compat-ible with a sustainable and more equitable world. At stake is long-term human well-being. "If we come to understand the human situation deeply, comprehensively, and clearly, then what needs to be done—both in our personal lives and the world around us—becomes clear."

Toward this end, Macdonald advocates development of *deep understanding*, a variety of wisdom in which we integrate broadly based contextual knowledge with self-knowledge. Each of the fifteen chapters, divided in four parts, is considered a "Matter of Consequence": 1) Big Picture Reality: the nature of primal reality, the development of complexity, understanding human mentality, the question of cosmic purpose; 2) Humanity's

Contextual Reality: sociocultural context, economic context, biospheric context; 3) Personal Reality: self-knowledge, freedom and responsibility, developing deep understanding, significant doing; 4) The Future: predicting the future, creating the future, the 2050 vision, and doing what needs to be done.

The Year 2050 Vision, inspired by the perennial philosophy, Spinoza, Ken Wilber, Ervin Laszlo, Buckminster Fuller, and others, describes a gradual transition from the twentieth-century high-consumption world to a sustainable, economically functional, and politically stable world in which everyone has an adequate standard of living: 1) Physical Sustainability: meets the needs of the present without compromising the ability of future generations to meet their needs (a May 2003 Google check of the Web located 2,150,000 Web pages with the word *sustainability*, up from 763,000 in May 2001; "clearly, something massive is happening"); 2) Universal Provisioning and Economic Stability: the economy redesigned to provide for the human population, rather than make a lot of money for a few people; energy needs are largely met by captured solar energy; 3) Work and Leisure: every able person is expected to spend a certain amount of time in socially relevant activity, and the essential work of society gets done; 4) Political Stability: attending to everyone's basic needs has greatly enhanced political stability; 5) Community and Civic Culture: most people are involved with their local community and a geography-irrelevant community of common interests, and honor both local and world culture; 6) Inner Development and Transformation: academic learning, skill development, ethical development, spiritual maturation, and development of creativity are considered aspects of a comprehensive process.

[NOTE: A very readable integration of many future-thinkers, with an appendix of relevant groups and resources. Although "many people" may now have "awareness of these realities," the number is still far from enough for decisive political action.]

Annotated Bibliography (O–Z)*

Only a few of the many books dealing with the future can be listed here. Selection criteria include recency of publication, trustworthiness of the author, readability, and usefulness to nonspecialists. Certain classic works have also been included, as well as a few other books not fully meeting the selection criteria but still likely to be of interest. Information on the most recent books is available on the World Future Society's Web site (wfs.org), which also provides information through Future Survey, *a monthly newsletter describing and commenting on the latest books and articles dealing with the future and major public-policy issues.* Future Survey *is the best available guide to current literature dealing with the future and is available by subscription.*

* Part 4 of four parts, one in each volume in this series. Reprinted with permission from Edward Cornish and the World Future Society.

Ogilvy, James A. *Creating Better Futures: Scenario Planning as a Tool for a Better Tomorrow.* Foreword by Peter Schwartz. New York: Oxford University Press, 2002.

A co-founder of the Global Business Network argues that we "suffer from a lack of sufficient idealism." Popular images of the future tend to be grim. Ogilvy takes "an unabashedly hopeful" view but says there is nothing inevitable about better futures. We must create them, and we can help people do so by framing alternative scenarios customized to their particular situations. This is not a how-to book for scenario planning but rather a general discussion of its usefulness in improving organizational futures.

Organisation for Economic Cooperation and Development. *21st Century Technologies: Promises and Perils of a Dynamic Future.* Paris: OECD, 1998.

The OECD's secretariat prepared this report for the World

Exposition in Hannover, Germany, in 2000. The report consists of seven articles by experts on topics like biotechnology and the macro conditions conducive to realizing technology's potential.

Orwell, George. *Nineteen Eighty-four*. New York: Harcourt, Brace and Company Inc., 1949.

A classic dystopian novel. The atomic wars Orwell refers to as occurring in the 1950s never happened, but his portrayal of a totalitarian state remains of lasting interest. The book can best be viewed as a satire rather than a serious attempt to forecast the future. Orwell wanted to attack the totalitarian tendencies by means of a satire, so he showed them in exaggerated forms. But the tendencies remain alive and well today, and may be seen in such phenomena as the vilification of enemies, the terrorization of non-conformists, and the rewriting of history to fit current political dogmas.

Pearson, Ian, ed. *The Macmillan Atlas of the Future*. New York: Macmillan Reference, 1998.

Full-color maps and graphics provide a vivid overview of where we are headed in the new millennium. An international team of leading analysts predicts developments in such areas as space exploration, economics, life expectancy, biodiversity, democracy, and more. Though dated, the book's approach is eye-catching and still interesting.

———, and Chris Winter. *Where's IT Going?* New York: Thames & Hudson, 1999.

Two British telecommunications experts offer a forecast for future developments in information technology. Specific topics include computers, communications, financial systems, and social institutions and behavior.

Petersen, John L. *Out of the Blue: Wild Cards and Other Big Future Surprises*. Lanham, Md.: Madison, 1997.

John L. Petersen, president of the Arlington Institute in

Arlington, Virginia, examines the potential impacts of such "wild card" events as an asteroid collision, the collapse of the U.S. dollar, a shift in the Earth's axis, and the perfection of techniques for cloning humans. This rapid ride through scores of scenarios is designed to get you to think "out of the box" and learn how to manage surprises.

Pirages, Dennis C., ed. *Building Sustainable Societies: A Blueprint for a Post-Industrial World*. Armonk, N.Y.: M.E. Sharpe, 1996.

This collection of twenty previously unpublished essays by noted scholars analyzes the implications of economic development as part of the search for an economic system that can be sustained over time.

————, and Theresa Manley De Geest. *Ecological Security: An Evolutionary Perspective on Globalization*. Lanham, Md.: Rowman & Littlefield Publishers, 2003.

Pirages is a professor of international environmental politics at the University of Maryland with years of distinguished scholarship and numerous books to his credit. This book is a well-organized, systematic discussion of the ecological issues of the near-term future, with lucid analyses and many suggestions for solving the problems of a globalizing world.

Polak, Fred L. *The Image of the Future*. New York: Elsevier, 1973.

A Dutch scholar prepared this study of the role of images of the future down through history. Translated by Elise Boulding, a sociologist and futurist.

Prantzos, Nikos. *Our Cosmic Future: Humanity's Fate in the Universe*. Cambridge, U.K.: Cambridge University Press, 2000.

Billions of years from now, the Sun will run out of fuel and grow dark, making the Earth uninhabitable. Prantzos, and engineer with the Swiss Technical Institute in Zurich, offers an escape plan for humanity—in plenty of time to prepare for the crisis.

Prehoda, Robert W. *Designing the Future: The Role of Technological Forecasting*. Philadelphia, Pa.: Chilton Books, 1967.

Technological forecaster Robert Prehoda presents a rationale for man's potential ability to foresee accurately the future capabilities and results of applied science. The book defines technological forecasting as "the description or prediction of a foreseeable invention, specific scientific refinement, or likely scientific discovery that promises to serve some useful function." Prehoda describes several approaches to technological forecasting; his primary technique is through what he calls "the Hahn-Strassmann point." The name comes from the Hahn-Strassmann experiments in 1938, which showed the possibility of uranium fission. His method is to look for analogous situations; i.e., laboratory achievements that show the possibility of some major advance on a practical scale, then forecast the practical results that could be based on this laboratory achievement.

Rees, Martin. *Our Final Hour*. New York: Basic Books, 2003.

Rees, the Astronomer Royal of the United Kingdom, takes time from his cosmological work to issue a bleak warning to earthlings: "I think the odds are no better than 50-50 that our present civilization on Earth will survive to the end of [this] century." He offers a long list of worries, from black holes to unstable individual buildings, but he also suggests reasons why many of his horrors may not prove as horrible as he fears.

Reibnitz, Ute von. *Scenario Techniques*. New York: McGraw-Hill, 1987.

Scenario techniques are suitable for all projects dealing with complex, interrelated problems. This book clearly shows how scenarios may be used in strategic planning, individual planning in organizational departments, and personal planning.

Renesch, John E. *Getting to the Better Future: A Matter of Conscious Choosing*. Foreword by Anita Roddick. San Francisco, Calif.: New Business Books, 2000.

Humans have hardly reached full maturity as a species, the author contends, as evidenced by conditions throughout the world. The present course suggests that the future of our world will unfold by default—the result of haphazard and random decisions based on short-term expediency. However, the author offers an alternative future—one created by conscious intention with the longer term in view.

Rescher, Nicholas. *Predicting the Future: An Introduction to the Theory of Forecasting*. Albany, N.Y.: State University of New York Press, 1998.

The author, a professor of philosophy at the University of Pittsburgh, provides a thorough discussion of the nature and problems of prediction. The volume discusses the ontology and epistemology of the future, predictive methods, evaluation of predictions and predictors, obstacles to prediction, and other aspects of prediction. Rescher also focuses on the theoretical and methodological issues of prediction, but does not try to make predictions of his own for the future.

Ringland, Gill. *Scenario Planning*. Chichester, U.K.: Wiley, 1998.

A comprehensive guide to using scenarios in business, it can function well both as a general primer and as a detailed textbook. The author, an executive with the London-based information technology firm ICL, notes that people in business sometimes confuse scenarios as forecasts. Rather, they are ways to understand the total environment in which business operates.

Rischard, Jean-François. *High Noon: 20 Global Problems, 20 Years to Solve Them*. New York: Perseus Books, 2002.

A vice president of the World Bank argues that the current ways of dealing with complex global issues are not up to the job. He argues that "each global issue should have its own problem-solving vehicle" and proposes twenty global issues networks.

Rubenstein, Herb, and Tony Grundy: *Breakthrough Inc.: High Growth Strategies for Entrepreneurial Organizations.* Harlow, U.K.: Financial Times/Prentice Hall Pearson Education Ltd., 1999.

This book by two business consultants, focusing on how companies can grow rapidly, discusses trend analysis and other futurist techniques.

Sale, Kirkpatrick. *Rebels against the Future: The Luddites and Their War on the Industrial Revolution.* New York: Addison-Wesley, 1995.

A long-time critic of technology recounts the famous Luddite uprising against the new manufacturing equipment powered by steam or water engines in the 1790s. The British Parliament dispatched an army of 14,000 men to put down the rebellion. Viewing the Luddites as martyrs to the cause of halting the technological juggernaut, Sale proposes a wholesale dismantling of industrial technology as we know it.

Salmon, Robert. *The Future of Management: All Roads Lead to Man.* Translated by Larry Cohen. Oxford, U.K.: Blackwell Publishers, 1996.

Salmon, the former vice chairman of the cosmetics giant L'Oreal, offers business executives a visionary guide to the coming decades. Salmon believes a new economic order is emerging on the basis of human dynamics and aspirations rather than short-term financial gains achieved at employees' and customers' expense. It is devotion to human potential that will unlock the future, Salmon believes.

Schnaars, Steven. *Megamistakes: Forecasting and the Myth of Rapid Technological Change.* New York: Free Press, 1989.

A marketing professor at the City University of New York

argues that 80 percent of business forecasting has been dead wrong. His analysis focuses on the last three decades in which entrepreneurs, managers, and forecasters have repeatedly anticipated great success for such developments as the videophone only to have them fail in the marketplace. In this entertaining volume, Schnaars explores the question of why consumers repeatedly reject the new products they are offered, thereby invalidating the forecasts of the developers.

Schrage, Michael. *Serious Play: How the World's Best Companies Simulate to Innovate*. Foreword by Tom Peters. Boston, Mass.: Harvard Business School Press, 1999.

A discussion of business innovation by means of models, simulations, gaming, and prototyping.

Schwartz, Peter. *The Art of the Long View: Planning for the Future in an Uncertain World*. New York: Doubleday/Currency, 1991.

Schwartz, a former leader of the Global Business Network, shows how composing and using scenarios can help people visualize and prepare for a better future. Oriented largely toward business rather than individuals.

———, Peter Leyden, and Joel Hyatt. *The Long Boom: A Vision for the Coming Age of Prosperity*. Reading, Mass.: Perseus Books, 1999.

An optimistic vision of the first two decades of the twenty-first century.

Sherden, William A. *The Fortune Sellers: The Big Business of Buying and Selling Predictions*. New York: Wiley, 1998.

Sherden launches a frontal assault on futurists, economists, stock market gurus, weather forecasters, technology prophets, and others who make forecasts for money. The gravamen of his indictment is that forecasters are frauds because the future is unknowable: When forecasters do manage to get something right, it's simply by chance. This is a highly readable exposé but so one-sided that it's hard to take seriously. Futurists do not need to be told that forecasts often prove wrong, but they may find it useful to hear what a critic has to say.

Shostak, Arthur B., ed. *Viable Utopian Ideas: Shaping a Better World*. Armonk, N.Y.: M.E. Sharpe, 2003.

A collection of forty-seven original essays exploring pragmatic reform possibilities of special interest to futurists. It features ideas by such leading forecasters as Wendell Bell, Joseph F. Coates, Lane Jennings, Michael Marien, and many others.

Slaughter, Richard A. *The Foresight Principle: Cultural Recovery in the 21st Century*. Westport, Conn.: Praeger, 1995.

Why is foresight useful? How much does it really cost, and who should support it? What are the real megatrends?

———. *The Knowledge Base of Futures Studies*. 3 vols. Hawthorn, Victoria, Australia: DDM Media Group, 1996.

This three-volume set provides an in-depth, authoritative, and truly international overview of futures studies. Volume 1, *Foundations*, considers the origins of futures studies and discusses some of the social, cultural, and historical reasons for their emergence. Volume 2, *Organizations, Practices, Products*, begins with case studies of five very different futures organizations from different continents. Volume 3, *Directions and Outlooks*, describes recent developments and innovations in futures studies itself. This set of volumes will aid anyone working in any of the emerging futures professions. *Note:* A fourth volume is included in the CD-ROM version.

Stableford, Brian, and David Langford. *The Third Millennium: A History of the World: AD 2000–3000*. New York: Knopf, 1985.

Fanciful scenarios for possible developments in the centuries ahead.

Stewart, Hugh B. *Recollecting the Future: A View of Business, Technology, and Innovation in the Next 30 Years*. Homewood, Ill.: Dow Jones-Irwin, 1989.

Where growth is involved, scientist/engineer Stewart believes it is possible to "recollect" the future—that is, apply laws of growth similar to those in biology to the development of new industries, energy use, and the economy. Stewart believes that if his arguments are correct we will see the dawn of an important and new industrial, energy-use, and economic surge beginning before 2000. The book should appeal strongly to people interested in ways to anticipate future technology.

Stock, Gregory. *Redesigning Humans: Our Inevitable Genetic Future*. New York: Houghton Mifflin, 2002.

An expert on recent advances in reproductive biology discusses the ethical dilemmas occurring as we gain the ability to choose our offspring's genes. Biological enhancements of human abilities may challenge what it means to be human. The author has served as director of UCLA's Program on Medicine, Technology, and Society. Compare with Fukuyama's *Our Posthuman Future*.

Tenner, Edward. *Why Things Bite Back: Technology and the Revenge of Unintended Consequences*. New York: Alfred A. Knopf, 1996.

New technologies are likely to produce "revenge effects"—bad consequences that were not intended. Advanced safety systems can lead to overconfidence, as in the case of the *Titanic* sinking. This popularized, largely anecdotal discussion also notes that technology's "bites" have often had good effects: The sinking of the *Titanic* led to new measures to deal with the hazards of icebergs and ocean travel in general.

Thomson, Sir George. *The Foreseeable Future*. Rev. ed. Cambridge, U.K.: Cambridge University Press, 1960.

Sir George Thomson, the Nobel Prize-winning British

physicist, published this book in 1955. The author focused on sources of energy, transportation, communication, meteorology, natural resources, food production, and intellectual development. Two decades after his book appeared, a reviewer in *The Futurist* reported that Thomson's twenty-year-old forecasts had generally turned out to be remarkably accurate. Thomson had correctly foreseen the energy crisis in the 1970s and the triumph of the computer.

Toffler, Alvin. *Future Shock*. New York: Random House, 1970.

This international best-seller argues that increasing numbers of people are suffering from the impact of too rapid social change. Toffler believes that the problem may become increasingly severe in the years to come. A final chapter, "The Strategy of Social Futurism," proposes a number of ways in which society can learn to cope with future shock. The author outlines his ideas about how to make democracy more anticipatory in its character.

————. *Powershift: Knowledge, Wealth, and Violence at the Edge of the 21st Century*. New York: Bantam, 1990.

Powershift is the third and final volume of a trilogy that began with *Future Shock* (1970) and continued with *The Third Wave* (1980). Toffler examines three forms of power: 1) knowledge (information, communications, and media), 2) wealth (business, financial), and 3) violence (government, politics). He argues that power is now shifting from violence and wealth toward knowledge and provides a highly browsable compendium of fascinating anecdotes and insights.

————. *The Third Wave*. New York: William Morrow, 1980.

The author describes a new civilization that he believes is emerging from our present industrial civilization, but the best part of the book may be the author's ability to provide numerous fascinating insights, useful perspectives, and plausible anticipations concerning current trends.

————, and Heidi Toffler. *War and Anti-War: Survival at the*

Dawn of the 21st Century. Boston, Mass.: Little, Brown, 1993.

A readable but disturbing discussion of what war may be like in the future, including widely dispersed nuclear weapons available to small groups such as Asian warlords or Mafia families. A nuclear bomb may explode in Washington or other big city without anyone knowing who is responsible.

Vision Center for Futures Creation. *A Tale of the Future*. E-book. Göteborg, Sweden: Visionscentret Framtidsbygget, 1998.

This electronic book offers an extended scenario or "social science fiction," based on a study conducted by a group of twenty-six young Swedes. The electronic book is available in Swedish or English.

Wagar, W. Warren. *Good Tidings: The Belief in Progress from Darwin to Marcuse*. Bloomington, Ind.: Indiana University Press, 1972.

The belief in progress is sometimes viewed as "the religion of modern man." In this book, Wagar analyzes how this belief changed from 1880 to 1970. A good companion volume for J.B. Bury's earlier work *The Idea of Progress*.

———. *The Next Three Futures: Paradigms of Things to Come*. Westport, Conn.: Praeger, 1991.

Wagar, a historian and futurist based at the State University of New York in Binghamton, identifies three major camps of futurist thinking: technoliberals, radicals, and counterculturalists. The book as a whole provides a vital introduction to the diversified field of futurist inquiry.

———. *A Short History of the Future*. 3rd ed. Chicago, Ill.: University of Chicago Press, 1999.

Wagar offers a detailed scenario for world developments during the twenty-first century.

Weiner, Edith, and Arnold Brown. *Insider's Guide to the Future*. New York: Boardroom Books, 1997.

Two veteran futurists specializing in analyzing trends for business clients offer insights into the new "Emotile Society," which blends emotions and mobility. Knowledge will be the greatest economic asset, but it will be limited by time: Information that is incredibly valuable one moment may be worthless the next.

Weisbord, Marvin R., and Sandra Janoff. *Future Search: An Action Guide to Finding Common Ground in Organizations and Communities.* San Francisco, Calif.: Berrett-Koehler, 1995.

The future-search process focuses on resolving conflicts, generating commitment to common goals, and taking responsibility for action. This practical guide offers techniques for running successful future-search conferences.

Weldon, Lynn L. *The Future: Important Choices.* Boulder, Colo.: University Press of Colorado, 1995.

This introductory text for college-level courses in futures studies is also useful for general readers, offering a succinct summary of major world issues and differing views.

Wilson, Edward O. *The Future of Life.* New York: Knopf, 2002.

Wilson, a Harvard biology professor, warns of an "armageddon" due to humanity destroying the Earth's living environment.

World Future Society. *The Futurist Directory: A Guide to Individuals Who Write, Speak, or Consult About the Future.* Bethesda, Md.: World Future Society, 2000.

A listing of nearly 1,400 people that includes specialization, employment, publications, street and electronic addresses, and telephone numbers. Geographical and subject indexes.

Worldwatch Institute. *The State of the World.* New York: W.W. Norton. Published annually.

This annual volume has been published every year since 1984. Lester R. Brown, the institute's founder and first president, led the development of these reports, and their readability and high quality has continued under the leadership of the institute's new president, Christopher Flavin. The reports offer authoritative information on current world trends, with special emphasis on the environment. Readable and trustworthy information. Specific topics vary from year to year.

Worzel, Richard. *The Next Twenty Years of Your Life: A Personal Guide into the Year 2017*. Toronto: Stoddart, 1997.

Clear, concise discussion of how technologies and other changes will affect your family, your work, your health, and your lifestyle. Among Worzel's forecasts: the end of retirement, the end of television, the biggest stock market boom in history, and life spans extended beyond age 100.

Zey, Michael G. *The Future Factor: The Five Forces Transforming Our Lives and Shaping Human Destiny*. New York: McGraw-Hill, 2000.

A general discussion of the human future, with emphasis on the new possibilities opened by emerging technologies. Topics include nanotechnology, cloning, genetic engineering, smart machines, space settlement, and general human progress. The optimism of this book's view of the human future is extreme: Probably no previous book has suggested that humans might someday save the universe from eventual death in a "Big Chill" or "Big Crunch," as augured by cosmology's Big Bang theory.

* This bibliography is accessible online and continuously updated. Please visit *Futuring* at wfs.org/futuring.htm.

Notes on Contributors

Jan Amkreutz has spent his professional life in information technology, as a programmer, research professor, executive, and entrepreneur. A Dutch national, he has lived in Europe, the Far East, Canada, and the United States. He is the author of *Digital Spirit: Minding the Future*, and president of Digital CrossRoads, a digital technology consulting practice. His Web site: www.digeality.com.

John A. Blackwell is the owner of The Creative Futures Center, which is dedicated to the participation of individuals in the creation of their own future. He combines his Royal Dutch Shell background in global team communications and facilitation with an M.S. in studies of the future to offer a range of services for individuals and teams.

John Cashman is a futurist with Social Technologies, LLC, in Washington, D.C. His interests lie in identifying and interpreting emerging forces that change how people live their lives. He is the director of a multisponsor study on global lifestyles. He can be reached at john.cashman@soctech.com.

Thomas Gangale is an aerospace engineer and a former U.S. Air Force officer. He is the executive director of OPS-Alaska, a think tank based in Petaluma, California, and a graduate student in international relations at San Francisco State University. He is the author of the California Plan to reform the presidential nomination process. His writings are available on the OPS-Alaska Web site at www.ops-alaska.com.

Phil Gyford is a Web site designer and programmer based in London, England. He has an M.S. in futures studies from the University of Houston–Clear Lake and runs a variety of Web sites based on the past and future, which can be found at www.gyford.com.

Glenn Hough has a master's degree in studies of the future. He came to his master's work via the path of creative writing in science fiction, having written three novels and several movie scripts. In future studies, he has found an arena in which to express the sentiment: Good conscience before blind obedience to the tenets of society.

Melvin Konner, Ph.D., is the Samuel Candler Dobbs Professor of Anthropology at Emory University in Atlanta. He is the author of *The Tangled Wing: Biological Constraints on the Human Spirit, revised edition; Becoming a Doctor: A Journey of Initiation in Medical School;* and *Unsettled: An Anthropology of the Jews*. He is interested in how our evolutionary past and our biological heritage shape our possibilities for the future and in how biological and medical discoveries are changing our lives.

Jeff Krukin has an M.S. in studies of the future and has been an information technology professional since 1981. He is passionate about the human-space connection and commercial space development, and has been writing and speaking about this for more than twenty years. His material has appeared in *Space News*, the *Houston Chronicle, Houston Business Journal*, and other publications, and he has spoken at numerous conferences. He is chairman of ProSpace, a space advocacy organization, and is launching a new career as a space speaker, writer, and analyst. Please visit www.jeffkrukin.com to learn more about space and contact the author.

Graham T.T. Molitor is president of Public Policy Forecasting, and vice president and legal counsel of the World Future Society. He headed lobbying staffs at General Mills and Nabisco, chaired a legislative Commission on the Future, directed research for the White House Conference on the Industrial World Ahead, served on the White House Social Indicators Committee, headed research for both of Vice President Nelson Rockefeller's presidential campaigns as well as had part-time roles in two other presidential campaigns, worked as legal counsel in the U.S. Congress, and served with the assistant chief of staff at the Pentagon. Publications include *The Power to Change the World: The Art of Forecasting*, 2003; *The 21st Century* (coeditor), 1999; and *The Encyclopedia of the Future* (coeditor and editorial board chairman), 1996.

Alexandra Montgomery is a graduate of the University of Houston–Clear Lake Studies of the Future program, with professional interests in futures studies, and community and family futures. Alexandra is a freelance futures researcher in Houston and a member of the World Futures Studies Federation and the World Future Society.

Jim Pinto was founder and formerly CEO of a high-technology company based in San Diego, California. He is now a technology futurist, angel investor, speaker, writer, commentator, and consultant. His recent book, *Automation Unplugged*, was published by ISA. He invites your feedback, ideas, suggestions, and encouragement. Visit his Web site at www.JimPinto.com; or E-mail: jim@jimpinto.com.

Arthur B. Shostak, Ph.D., (Editor) holds the title of emeritus professor of sociology after recently retiring from Drexel University (Philadelphia, Pennsylvania), where he had been a professor since 1967. Since he began college teaching in 1961, he has specialized in trying to apply sociology to real-time problems ("challenges") and in shaping and communicating long-range forecasts. While at Drexel, he directed a two-year study of teenage attitudes toward the world of work and related matters. He has written, edited, and coedited more than thirty books and more than 160 articles, and was presented with the Pennsylvania Sociological Society's Distinguished Sociologist Award in 2004. He especially recommends his 2003 edited collection, *Viable Utopian Ideas: Shaping a Better World* (M.E. Sharpe, Armonk, N.Y.). You can contact him at shostaka@drexel.edu.

John Smart is a developmental systems theorist who studies accelerating change, computational autonomy, and a topic known in futurist circles as the technological singularity (see http://SingularityWatch.com). He is president of the Institute for the Study of Accelerating Change (http://Accelerating.org), a nonprofit community that promotes awareness, analysis, and selected advocacy of communities and technologies of accelerating change. He produces the Accelerating Change Conference, an annual meeting of three hundred leading thinkers and students at Stanford University, and edits ISAC's free newsletter, *Accelerating Times*. John lives in Los Angeles and can be reached at johnsmart@accelerating.org.

Sarah Stephen is working toward a master's degree in applied communication at Royal Roads University in British Columbia. She wrote "Letters to Unborn Daughters ..." as a student in Lynn Burton's future studies class at Simon Fraser University. Her e-mail address is sarahmstephen@yahoo.ca.

Dana Wilkerson-Wyche is a native Houstonian who took an interest in law enforcement while serving in the U.S. Marine Corps. She attended the University of Houston, where she received a B.A. in Criminal Justice, with a minor in Psychology. She is currently studying Criminology at the University of Houston–Clear Lake.

Index

Abortion, 48

Advocacy groups and genetics, 177

"Aerofoils and Aerofoil Structural Combinations," (Gorrell/Martin), 112

Afghanistan, 156

African-Americans perspectives on genetic testing, 177

AIDS, 78

Albright, Madeleine, 151

Alcohol, 75

Aldrin, Buzz, 181

American pioneers, 32

Amino acids, 68

Ancient civilization, 150
   mass extermination, 151

Andromeda Strain, The, (Crichton), 92

Animals
   and genetic modification, 76–77

Annals of Earth (Fousek), 184

Annan, Kofi, 151

Antidepressants, 74–75

Apollo missions, 102, 111, 124

Architecture of Global Governance, The: An Introduction to the Study of International Organizations (Muldoon), 185

Artificial insemination, 47

Artificial Intelligence, 175–176

Artificial Life: The Coming Evolution (Farmer/Belin), 92

Artificial organisms, 92

Associated Press, 105

Asteroids, 104–105

Astrobiologist, 38

Astronomy, 79

Atom, 44
   on building structures, 85–86

Attention deficit disorder, 75

Autonomous intelligences (AIs), 39, 40

Bainbridge, William, 40

Balkans, 156

Bangladesh, 80

Barnes, Stephen, 100

Battleship Row, 126

Belin, Alletta d'A., 92

Bell, James John, 13

Bible, 149

Bigelow Aerospace, 181

Bioanalysis, 83

Biodiversity, 50

Bioinformatics, 46

Biological systems, 39

Biology, 37, 40, 87

Bio-Molecular Revolution, 44, 45

Bioreactor food production
   cloning only desirable crop components, 68–69

Biotech advances, 83
   designer babies, 46–47
   sweeteners, 67–68

Body parts
   and cloning of, 50
   on regenerating, 50

"Book of Life," 51

Bostrom, Nick, 175

Brave New World (Huxley), 50

Bread, 69

Brown, Louise, 47, 78

Bush, George W., 102, 153
   and budget for space, 115
   his space plan, 105, 110

Business Week, 178
   on nanotechnology, 87

California Institute of Technology, 85

Capitalism, 14, 97

Carbon nanotubes, 88

Carter, Jimmy
   on biotech, 70

Chemistry, 87

China, 80, 111, 118–119

Chlorofluorocarbons (CFCs), 154

Cloning, 46, 47, 79

of animals, 49–50
of humans, 50
involves genetic procedures, 49
organ transplants, 50
Cognitive science, 83
Coicaud, Jean-Marc, 183
Cold War, 102, 180, 185–186
Colonial Americans, 32
Commercialization of genetic technologies, 178
Communism, 121, 154, 186
Complex molecules, 37
Computational optima, 37
Computer chip
for the brain, 74
Computers, 34, 44
challenges of, 17
Computer science, 87
Condom, 77
Congenital thyroid disease, 48
"Cosmic Calendar," 33
Cosmic Discovery (Harwit), 34
Cosmos I, 181
Crichton, Michael, 92
Cultural Creatives, The (Ray), 187
Culture Shift in Advanced Industrial Society (Inglehart), 31

Darwin, 37–38
Dator, James, 95
Dator's Law for Future Studies (Dator), 95
Dealey Plaza, 126
Dean, Howard, 17
Deforestation, 151
Democracy, 154, 156
Developmental physics, 35
and hidden from view, 39
Diabetes, 77
Diet, 75
Digital future
imaginary day, 19
importance of, 29
Inter Society News, 20

special shirt of the future, 21
Digital twin (DT), 21
Disease, 72–75
elimination of, 47
DNA, 21, 37, 38, 51, 83
possibility of editing, 86
Dolly (cloned sheep), 49
Double-Edged Helix, The: Social Implications of Genetics in a Diverse Society (Alper) 176–177
Down syndrome, 48
Doyle, Michael W., 183
Dream of Space Fight, The (Wachhorst), 112
Drexler, Eric, 38, 92, 96
and consequences of technology, 90
on stacking atoms, 86
Drugs, 47
Dyson, Freeman, 91

Earth
life on, 80
Earth-orbit transportation infrastructure, 106
Eating disorders, 75
eBay, 17
Echo-boom, 152
Economics, 84, 98
Eddie Bauer, 90
Ehricke, Kraft, 104
Einstein, Albert, 95
on questioning, 147
Embryos
and genetic disease, 78
Engines of Creation (Drexler), 38, 86, 96
Essential Institution for the 21st Century, An (Mendlovitz), 184
"Ethnic cleansing," 49
Eugenics, 48
on cause and effect, 49
Europa, 80

Europe
on GM foods, 70
Evils
fear, 96
ignorance, 96
want, 96
Evolution, 33–34, 37, 149
Evolutionary Development
(Evo-Devo), 33–34, 38
Extinction, 91

Farmer, J. Doyne, 92
Feynman, Richard, 85
Food
and biotech, 66–67
curbing insect damage, 67
Fossil fuels, 152
Fousek, John, 184
Freedom, 157
Fuller, Buckminster, 188
Fuller, Steve, 176
Future Needs Us, The
(Dyson), 91
Future Shock (Toffler), 83
Future Survey (newsletter), 175
Futurist, The (Bell), 13
Futuristics, 147

Gagnon, Bruce K., 121
Gardner, Anne-Marie, 183
Gates, Bill, 31
Gattaca (movie), 48
Gehrig, Lou, 48
Gender and genetics, 177
Genes, 44
science of, 157
Genetically modified foods
(GM), 44, 76
on "libraries" of GM
bacteria, 77
and mixed response, 70
Genetic "blueprints," 51
Genetic complexity in human
disease and behavior, 177
Genetic crop modification,
66–67

Genetic discrimination, 178
Genetic engineering, 38–39, 46,
47
implications of, 52–57
Geneticists in society, 177
Genetics, 46, 72–73, 178
in human disease, 177
and job opportunities, 51
in society, 177
Genetic screening, 48
Genetic Screening Study
Group, 177
Genocide, 151
Genocide Convention, 184
George Washington University
Space Policy Institute, 182
Germany, 156
Germ-line therapies, 47, 48
Germ theory, 77
Gibson, Rowan, 17
Global Governance, 185–187
*Globalization of Human Rights,
The* (Coicaud/Doyle/
Gardner), 183
Global Network Against
Weapons and Nuclear
Power in Space, 121
Global War, 104
Global Warming, 151
Goddard, Robert, 180
"Golden rice," 69
Gorbachev, Mikhail, 151
Gorrell, E.S., 112
Grapes, 76
Growth hormone, 74

Harwit, Martin, 34
Hawking, Stephen, 48
Health care
on anesthetics, 46
and natural forces, 46
on vaccines, 46
Heinlein, Robert A., 127
Hitler, 49, 154–155
Hudgins, Edward L., 180
Human genetic code, 44

Human genome, 73
Human Genome Diversity
  Project, 177–178
Humanity
  where we came from, 40
Human rights
  addressing demands of,
    183–184
Humans
  evolution of, 149–150
  on farming, 150
  kinds of control, 150
  and spiritual advance, 150
"Human-Space Connection,"
  113
Hussein, Saddam, 154–155
Huxley, Aldous, 50

Imaging, 72–73
Imagining a new world, 152–153
Immortality, 49
Implants
  for the deaf, 74
India, 118–119
Info-space, 37
Infotech, 83
Inglehart, Ronald, 31
Inner space, 35, 38
International Crime Watch
  needs to be established, 185
International Criminal Court,
  184
International Space Station, 100,
  106, 181
Internet, 17, 86, 148
In utero surgery, 48
In Vitro Fertilization (IVF), 78
Islamic Fundamentalists, 104
Italy, 80

Japan, 156
Jenkins, Anne, 176
Jones, Ron, 181
Joy, Bill, 91
Jupiter, 80
Jurassic Park (Crichton), 92

Kaku, Michio, 44, 121
Kapoor, Rakesh, 176
Kelly Aerospace, 181
Kennedy, John F., 128
  sending humans to the Moon,
    124–125
Kistler Aerospace, 181
Kistler K-1, 182
Kurzweil, Ray, 176

Lambright, W. Henry, 181
L-aspartyl-L-(Beta-cyclohexyl
  alaninc methyl ester), 67
Laszlo, Ervin, 188
Lee Jeans, 90
Leon, Ponce de
  "fountain of youth," 50
Letters to unborn daughters,
    (genetic engineering)
  2006, 52–53
  2031, 54–55
  2061, 55–56
  2091, 56–57
Life expectancy, 49
Linguistic User Interface (LUI),
  31, 32
L-morphine, 68
Lockheed Martin, 181
Logsdon, John M., 182
L-sugars, 68

MacDonald, Copthorne
  need of deep understanding,
    187–188
MacHale, 95
MacLeish, 127
Marien, Michael, 175
Mars, 14, 107, 111, 157
  desire to go to, 125
  most Earth-like, 126
  on sending humans, 124
  and water, 79–80
Mars Colony: 096
  and Bill of Rights, 145
  citizenship, 141

and committees, 134–136
crisis decision-making, 136–137
design principles, 143–145
and economy, 139–140
education, 142–143
as experiment, 132
feedback structure, 136
government of, 134
Judicial powers, 138–139
the Judiciary, 138
leadership, 142
on passage of a Bill, 137
Peace Corps, 139
rehabilitation procedures, 138
on social life, 140
and structure, 134
task allocation, 140
"Mars Direct Plan," 111
Mars rover, 100
Martin, H.S., 112
Mass extinction, 151
Matter
and self-replication, 86
Matters of Consequence: Creating a Meaningful Life and a World that Works (MacDonald), 187
Mead, Margaret, 155
Medicine, 72–75
Medieval economics, 98
Medieval times, 32
Memory chips, 88
Mendlovitz, Saul, 184
MEST compression, 33
Micro-electro-mechanical systems (MEMS), 88
"Mirror molecules," 68
Molitor, Graham T.T., 176
Moon, 126–127
how to use, 118–119
and military bases, 119, 120
Moore's Law, 175, 176
Muldoon, James P. Jr., 185
Multicelled developing embryo, 48

Mustard gas, 154–155
Mutually Assured Destruction (MAD), 117

Nanoclay particles, 89
Nano-electro-mechanical systems (NEMS), 88
Nanoelectronics, 87–88
Nanoscale engineering, 87
Nanoscale particles, 89
Nanostructured membranes, 89
Nanotechnology, 14, 38, 83
debate of, 90–91
effect on industry, 85, 91
eliminating want, 96
expectations of, 178–180
materials, 88
molecules working like machines, 74
and new jobs, 178
next revolution, 86
potential for, 87
products and services, 178
and self assembly, 86
Nano-Tex, 90
National Advisory Committee for Aeronautics (NACA), 112
National Aeronautics Space Administration (NASA), 68, 102, 106
adapting for the 21st century, 182–183
and budget, 115, 120
has controlled space, 180
and economic return, 110
helping the U.S. economy into orbit, 107
and lunar landings, 110
on private industry, 105–106, 111–112
Neurotechnology, 83
Neuroweapons, 83–84
Newberry, Deb, 178
New York Times, 96
Next Big Thing Is Really Small, The: How Nanotechnology

Will Change the Future of Your Business (Uldrich/Newberry), 178
Nixon, Richard
  and Mars, 124

Obesity, 77
Orbit-moon transportation infrastructure, 106
Organic chemistry, 37
Outer space, 34–35, 36
Ozone layer, 151

Parental tracking of children, (2020)
  the global picture, 63–64
  and global positioning system (GPS), 60
  manipulation of, 60–61
  the policy debate, 61–63
Pasteur, Louis, 68
Personal Digital Assistants (PDA), 31
Pharm-foods, 46, 47
  incorporate vaccines and therapeutic drugs into food staples, 69
Phenylketonuria, 48
Physics, 87
Planetary Organizations (POs), 22–23
Plenty of Room at the Bottom, There's, (speech), Feynman, 85
Polio, 154
Polymer "glue," 88
Population, 80, 151, 154, 155–156
Pregnancy, 78
Pre-natal diagnosis, 177
Prey (Crichton), 92
Princeton University, 91
Procter and Gamble, 89
Profit motive, 121
"Pyramid of Life," 35

Radio Shack rover, 181
Ravetz, J.R., 176
Ray, Paul H., 187
"Renewed Spirit of Discovery, A," (G.W. Bush), 105
  goal of, 116
  involving entire U.S. economy, 116
Reproduction, 149
Rethinking the Future (Gibson), 17
Rice, 69
Robotic arms, 74
Robots, 35
  on Mars, 125–126
Rubber manufacturing, 77
Russia, 119

Sagan, Carl, 33
Santa Fe Institute, 92
SAT (test), 74
Saturated fat, 77
Saturn, 80
Saving the world, 157–158
Scientific progress, 81
Scratchproof glass, 87
Segregation, 152
Self-restraint, 157
Sensor nets, 89
Sensors
  for the blind, 74
September 11, 151, 153
Sexually Transmitted Diseases (STDs), 77
Sickle-cell anemia, 48
Silent Revolution (Inglehart), 31
Singapore, 119
Singularity, 13, 18, 32, 40
Slavery, 154
Smallpox, 154
Solar arrays, 104
Sonar
  and seeing in the dark, 74
Space
  and career possibilities, 103–104, 110–111

211

and delays, 180–181
key to humanity's well-being, 104
more than a program, 102–103
on problems and possibilities, 182–183
in the 21st century, 181–183
Space Age, 100
SpaceDev, 181
Space: The Free-Market Frontier (Hudgins), 180
SpaceHab, 181
Space Island Group, 181
Space Policy in the 21st Century (Lambright), 181
Space Program, 102–103
 and conflict, 117
 as dangerous, 116
 new ideas for, 181
 on nuclear power, 117–118
Space shuttle, 106
Spinoza, 188
Spirit rover, 107
Sputnik, 181–182
Steinbeck, John, 97
Stem Cell, 46, 50
Steroids, 75
Stereoisomers, 68
Sun Microsystems, 91
Sunscreen, 89
"Super race," 49

Technology, 157
 artificial intelligence, 175–176
Tesla, Nikola, 125
Test-tube baby, 47, 78
Thermodynamics, 33
Third Great Revolution, 14
Tipping point, 155
 on knowledge, 152
 and will, 152
Titan, 80
Toffler, Alvin, 83
Tooth decay, 47
2050 (year)

vision of, 187–188
2035 (year)
 automated highway systems, 36
 biological you, 36–37
 Biology-Dominant Era, 30
 Digital Me (DM), 32
 electronic you, 36–67
 generation prime, 31
 hydrino tech, 33
 International Idea Network, 32
 and underground tubes, 36

Uldrich, Jack, 178
United Nations Constabulary, 184
 purpose of, 185
United States Department of Homeland Security, 89
United States Nanotechnology Initiative, 91
Universal Declaration of Human Rights, 1948, 183
Universe
 the big picture, 41
University of Hawaii at Manoa, 148
University of Houston–Clear Lake, 148

Verne, Jules, 72
Vietnam War, 152
Visions: How Science Will Revolutionize the 21st Century (Kaku), 44
Vitamin A, 69
Vitro fertilization, 47

Wachhorst, Wyn, 112
War on Terrorism, 89
Washington Post, 96
Water
 and war over, 151
Weight, 75
 and prescription drugs, 76

Wells, H.G., 125
Wetware, 39
When Machines Outsmart Humans (Bostrom),
Why the Future Doesn't Need Us (article) Joy, 91
Wilber, Ken, 188
Willpower, 76
Wilson tennis balls, 89
Wired (magazine), 96

Wisdom (MacHale), 95
World Future Society, 175
World Trade Center, 126
World War I, 154–155
Wright Brothers, 116, 180

Zapruder, Abraham, 126
Zubrin, Robert, 181, 182
    and "Mars Direct" Plan, 111

This book property of
Rector Public Library
Rector, AR 72461